S0-EKL-537

Vladimir Kolar

ISLANDS OF THE ADRIATIC

Published in the U.S.A. by
The Summerfield Press

Designed by
Miodrag Vartabedijan

Translated by
Karin Radovanović

Printed and bound in Italy by
SAGDOS, Brugherio, Milano

ISBN 0-8467-0465-X

Vladimir Kolar

ISLANDS OF THE ADRIATIC

The Summerfield Press
distributed by
The Two Continents Publishing Group

CONTENTS

An Invitation to Travel

For some time after the publisher kindly asked me to write this book, I cast about in my mind for the best approach, where to begin, how to identify the essential features of this region, this world that is in many ways unique and unusual.

The Adriatic islands.

The islands of the Yugoslav coast.

They belong to one sea and one country. They belong not only to the people who live there, but also to those who come occasionally to spend a well-earned holiday. For their inhabitants the islands are life itself: homeland, homestead, livelihood, tradition, customs, the joys and sorrows of everyday life. For visitors, the Adriatic islands are a place where sea meets land in a quiet, sunny embrace, where one's cares can be forgotten and new impressions received. Those for whom time is at a premium reach out eagerly for everything the islands have to offer: hot sunshine and fresh sea air, a glass of golden wine or a sun-ripened fig, the crystal peal of bells and the smell of drying seaweed, the pungent odor of olive oil and the white meat of grilled fish. These visitors reach out vigorously and unsparingly in the desire to take with them as much as possible. Those who stand at the thresholds of ancient stone houses remember a time when they gathered stone after stone to clear away the land, piling them onto low stone walls — a way of life that was hard, yet as bright and pure as the air they breathed. Painstakingly they worked tiny pockets of land to nourish the grapevines. They pinched the tiny buds from silver olive branches waiting patiently for the tree to attain its full growth, slowly and steadily, a centimetre or two every ten years. They set out across the sea and hauled in its yield.

The Adriatic is a small sea. The islands, too, are mostly very small. Even the largest, the monosyllabic Krk, is insignificant compared to island continents. The Adriatic Sea is in fact a large gulf between the Apennine and Balkan peninsulas. A long-distance swimmer might contemplate swimming the Adriatic, like the English Channel, from one shore to the other. The blue gulf thrusts deep, wedgelike, into the mainland of central Europe. It is of easy access by motorway from any part of Europe, at any time of year, the scenery and climate changing before one's eyes. In winter, for example, crossing the Alps over the vertiginous Saint-Gothard or the Simplon pass, or through the Mont Blanc tunnel, swooping into snowy valleys one thinks constantly ahead to one's destination: the murmur of the surf, and the merry dance of skipping waves, a fish flashing through the clear water, the sparkling reflection of sunlight on the seabed where algae sway in the eddying currents.

Or, as a contrast, take the summertime when caravans of cars stream south, towing boats on wheels, their dry prows thirsting for water. Chasing the sun, which moves on ahead of them, they never rest until their destination is reached: the tranquil cove of some island.

Again contrasts stand out — the damp blonde hair of a young vacationer soaking up the sun on a smooth rock overhanging the water, or the black headscarf of a peasant woman.

Life breathes in both these figures as they blend into a heterogeneous, many-nation crowd: native and visitor. Each is set against the other in contrasts that do not soften even when the dark sky is flecked with twinkling gold stars, when the hooting of a night owl and an embrace on a soft pine-needle carpet awaken pleasurable anticipation.

This small sea, an important trade route between Europe and the East, has been the scene of migration and conflict since the earliest times. If one takes as evidence legend and bits of terracotta that were once amphorae, the Greeks visited these shores when they were inhabited by the Illyrians. The wandering Greeks called them Lotus-eaters because they lived on the honeyed fruit of the lotus, which makes one forget all cares; or they thought they were cannibals of Cyclopean strength who, with animal instincts, tore apart the bodies of ill-fortuned wanderers. Out of the obscurity of the distant past come echoes of prehistoric life.

Behind stone walls we find ruins that bear witness to the coming of the Slavs and their settling on shores which were to become their homeland. War, persecution and epidemics punctuated decades of peaceful living and cultural expansion, its remains providing a graceful framework for present-day life on these shores — the patina on the facades of medieval palaces, stone fountains on ancient town squares, lacey church spires, ramparts and fortifications, towers and arsenals, tombstones, verses and epitaphs cut into stone, rose-windows and choir-stalls, altar paintings and golden icons by artists from the Adriatic and other shores, Franciscan and Benedictine monasteries, rare books and manuscripts, old maps and model ships, the sound of organ music swelling beneath Romanesque, Renaissance and Baroque arcades.

It seemed an impossible task to organise all this, to assign everything to its proper place, to distance oneself from the multitude of detail which so wonderfully blends into a whole, to encompass all that as a geographical unit makes up the "islands of the Adriatic."

A scientist from the Oceanography Institute of Split came to my aid. His approach to the sea and the islands, though different from mine, was no less human and possibly more essential. I absorbed his cold facts trying to give them human warmth.

If the surface of the sea were lowered a mere 90 metres, all the Adriatic islands would be joined to the mainland, he exclaimed, knowing full well that this was just theoretical speculation. And we both breathed a sigh of relief. For to be deprived of the islands would be a grievous loss for all. Owing to centuries of tectonic change there is a marked difference between the eastern and western coasts of the Adriatic. In the most recent geological era the eastern coast has sunk, and is probably still sinking. I looked at the scientist's intelligent face, at the maps on the walls of his study. I had the impression that we were sinking, so suggestive was his exposition. Simultaneously the western coast had risen with few indentations and no islands. I could see the coast of neighbouring Italy rising, a huge terrace overlooking the sea covered with brightly-coloured sunshades and white-jacketed waiters. The scientist just shook his head. He had serious objections to my literary extravagance. We were speaking of thousands of years; tectonic transformations do not comply with the improvising of a writer. We both concluded that for future generations on either coast there was no danger of any change.

But if the two coasts contrast so sharply, what about the seabed? Although we cannot see it, it has been thoroughly measured and examined. Its picture is unusual. If we were able, like astronauts, to view the sea from outer space, we would perceive that the bottom of the Adriatic resembles three gigantic steps descending ever deeper into the sea, from north to south, from the mouth of the Po River to the Straits of Otranto. The first step ends at about Ancona on the Italian coast and Pula at the tip of the Istrian peninsula. Here the Adriatic is shallowest – only fifty metres. The second step extends approximately halfway, somewhere near the island of Lastovo and the peninsula of Mt. Gargano on the Italian side. Here the depth averages about 130 metres. The third step runs to the tiny island of Palagruža. While you are rowing round that attractive, isolated isle, you never dream that at this point the sea bottom makes a sheer drop of 1228 metres into a dark depression, the deepest in the whole Adriatic.

The scientist at the Institute possessed all the facts. Like a magician pulling white doves out of a hat, he could easily furnish data for a portrait, a biography, a profile. If a sea could be said to have a temperament, one could describe the Adriatic as being calm, its surface disturbed occasionally by winds, of which there are several. The most treacherous blows from the southeast, driving huge breakers against the Yugoslav coast. The ancient Greeks called it a bad wind; sea-captains knew it as the Levanter. There are many other names for it, but all seamen know it is no joking matter. Strong enough to smash a concrete breakwater three metres wide, it can hurl rocks of several tons onto a beach. And it's no time to go swimming. Its name is often shortened to *jugo* or the south wind; it generally causes low spirits, and all quarrels are blamed on the cursed wind. The north wind, or *bura*, on the other hand, brings cool weather and drives the clouds south. Then everybody feels fine, singing along with the wind that whistles merrily over the chimneys and roofs.

The other winds fall under the category of breezes which are most appreciated by those in sailboats or on the beach in the scorching summer heat. Every afternoon before the sun starts its downward path the Adriatic is visited by a wind from the west called the *tramuntana*. Together with the *maestral*, a similar wind blowing from the southwest, they barely ripple the water near the coast, though out on the open sea they can be stronger. In the afternoon the sea sometimes becomes completely calm and smooth, a *bonaca* as it is called. Another warm breeze, known as the *bavica*, fans the water so gently it is scarcely noticeable. I prided myself on my new knowledge about the winds. All I needed was a sailboat so I could see in its billowing sails where the wind was coming from.

As we have said, the Adriatic is small, calm, and narrow if compared to other larger seas on our planet. One needs statistics, however, if one is to take exact measurements, to see how long it is, how wide, how deep. The scientist had no difficulty supplying me with figures. From the Straits of Otranto in the south to its northernmost point, the lagoon of Marano, the Adriatic measures 783 kilometres, or 423 sea miles, a mere trifle for ships that cross oceans. At its narrowest — from the point of Fažana in Istria to the mouth of the Po — it is a mere 103 kilometres, a distance that a speedboat can manage in a couple of hours. The crossing at the widest point — from the town of Vasto in Italy to Bar on the Yugoslav coast — comes to 355 kilometres, an all-night trip by ferryboat. All that remains is to measure the coastline. As the crow flies both coasts are approximately the same length. But if one measures all the twists and turns of the coastline, the figure comes to 3341 kilometres, of which the Yugoslav coast takes up 2092, and the Italian 1249. But enough said about figures.

Out at sea the straight line of the horizon is often broken by the hazy rounded outlines of an island. It appears to be hovering above the sea, its shores eaten away, misty and unreal. Is it a mirage such as a thirsty desert traveller might see? No, but something similar. As we approach an isolated island our gaze is exposed to the fragmentation of sunlight on the water. The scientific term is "refraction." In ancient times it gave rise to various fanciful and superstitious speculations. To sailors the islands appeared to be floating. Once when we were cruising about the Kornati archipelago, an angry squall blew up and set our boat to rocking; for a few unpleasant moments I thought the islands were swimming. Another time on a peaceful summer's night we were sailing in a channel between two islands, each marked by blinking lights from a lighthouse. And the two islands appeared to be sailing along with us.

The number of the islands appears to be infinite. They follow one another in rapid succession, some close to the coast, others farther out, some almost touching, others standing further apart. To count them would be the work of Sisyphus. My scientist, offering me a cup of tea, shook his head. These writers who think that islands can't be counted. They are all numbered, registered as if for a census, the information filed and noted in encyclopaedias. Reluctantly I returned to the figures. More than a thousand islands stretch along the eastern coast of the Adriatic — from the Soča River in the north to the Bojana River which separates Yugoslavia and Albania. On sixty-six of the 659 larger ones there are towns, fishing villages and hamlets; the remainder are largely uninhabited. In any case, there is an ample number to accommodate visitors wishing to spend their Adriatic holiday on an island retreat. Another 426 rocky islets rise from the water, inaccessible and without vegetation. There are also eighty-two reefs which may cause navigational headaches because they barely protrude from the water and are difficult to spot.

I thanked my interlocutor at the Oceanography Institute for the good cup of tea and the information which would help me to introduce the islands much as a person is represented by particulars such as date of birth, height and weight. At the door of his study I cast one more glance at the maps, the bookshelves and the big globe on his desk. Then it occurred to me to ask him how salty the sea was. I hesitated for a second thinking the question might be stupid, showing my ignorance. But the professor seemed to be expecting the question. We stood at the door another five minutes. He explained that the salt content of the Adriatic depended on many factors, that the water wasn't equally salty in all places. The salt content was highest farthest from the coast. Furthermore, the salt content was lowest in May and December when the rivers of the Adriatic watershed poured the most fresh water into the sea. And conversely, the sea's water is saltiest in February and September when the water level in the rivers is lowest. I was just about to shake hands and leave when I remembered another important question: why is the Adriatic so blue? When I write about the sea, the adjective "blue" is automatically prefixed, presumably suggested by all the cafés and restaurants along the coast called *The Blue Adriatic.* The professor smiled. The blue of the Adriatic has nothing to do with a glass of good Dalmatian wine, or a plate of grilled fish. The warmer and saltier the water, the bluer it appears, which explains why the sea is bluer in summer than in winter. The colour of the sea, which in different seasons combines infinite nuances of blue, grey and green, depends on the presence or absence of clouds. On dark cloudy days the sea is not as blue as on clear sunny days. Its colour is also affected by the colour of the seabed and the depth of the water. If the bottom is rocky, the sea is blue; if sandy, it tends towards a greenish hue. I'm not sure if painters are aware of this. Scientists further complicate matters by adding that the blueness of the Adriatic also depends on the amount of plankton and seaweed in the water, the position of the sun and angle from which the sea is viewed.

The professor watched me descend the steps, hardly believing I had run out of questions. But I had heard so much about the sea and the islands that I felt I needed a swim, regardless of the sea's salt content, colour, depth, breadth and whatever.

I returned to the everyday scene of island life... It was in the early hours of a summer morning, and a pearly dewdrop tumbled off a fig leaf. A donkey brayed from its stall. Hitching up her black skirt, a woman waded into the water. A squid, spread out and nailed to a post, was drying in the sun. A catch of silver-blue sardines wriggled in a tangle of nets. The air smelled of salted fish, asphalt, rosemary and the sea. A few swallows were chirping on a telephone line; from somewhere in the distance came the song of a peasant girl driving her flock to pasture, the words incomprehensible but the melody slow and sentimental. The donkey brayed again.

Now it was near the beach, tugging at the rope that tied it to a pillar. An old man with patched trousers hobbled painfully down the steps of his arbor. Cursing all the while, he picked up a stone and threw it at the donkey. A bunch of children laughed at the old man's helplessness. He threw a stone at them too, and they ran away down the beach, their thin laughter scattering like nuts on a stone floor. I was standing on the quay looking into the water. A starfish stretched out its points as if waking from a deep sleep. The sea was also awakening in the silence of a sunny morning. I threw a stone, and the rippling water carried away my picture.

These images kept returning, the whole atmosphere enveloping me. The night before gentle womanly fingers had eased the pain in my blistering shoulders. Through the open window we could hear a jazz band competing with the unvarying song of the crickets. A lighthouse at the end of the quay cast green shafts of light into the dark, enhancing the intimacy of an island night. All these images taken together, detail by detail, seemed to be something one could love infinitely and forever.

Then came the cars. On the small square, in big-city fashion, parking spaces were marked off. They started to sell the square, bit by bit, hour by hour. A solemn-faced fellow in uniform, a bag hanging from his shoulder, arrived to charge people for something that had once been free.

Fleeing before encroaching civilisation, I sailed to another island, putting as much water as possible between myself and the parking lot. And there I remained.

And this is exactly what I propose: leave your car on the mainland and take a steamer to a big island. There you'll find fishermen who can ferry you to a smaller island where you'll find peace and rest, sunshine and the sea — all in inexhaustible supply.

You'll also find yourself.

SUSAK

13. *The lonely island of Susak humps like a whale out of the morning sea. Isolated for many centuries Susak has developed, even more than the other Adriatic islands, its own special characteristics. The village, also called Susak, on the eastern end of this tiny island of sand has a population of 300.*

14. *Hills resembling desert sand dunes rest on a firm layer of limestone. The sandy soil supports grapevines and reeds, the only vegetation on the island.*

15. *The village with its steep, narrow streets perches at the edge of a sandy plateau. Gaining a livelihood mostly from their vineyards and fishing, the islanders rarely try to sell any of their produce, except to those who come to the island and ask. The two small restaurants on the island are of very recent date.*

16. *The women of Susak wear lightweight slippers and short skirts. Island life as a whole is geared to sand. To get around more easily the women carry everything on their heads. The reed fences along the roads act as windbreaks, the reeds thrusting long pliant roots deep into the sand hold it in place.*

17. *Cut off from the mainland and other islands, the inhabitants of Susak developed a distinct dialect, different customs and costume. They cherish unusual old traditions still practiced at weddings, during the grape harvest and on other occasions. The short skirt of the picturesque women's costume might well have inspired the mini-skirt.*

The Northern Kvarner Islands

CRES
The magic remains of humain endeavour

The islands of the Adriatic can be visited in any order. It matters little where you begin. You may catch your first glimpse of the sea at Split, or farther south at Dubrovnik, or from the shores of the Kvarner Gulf. Looking seaward you invariably discover an island, and often several; some are quite close, others form grey silhouettes against the horizon. Grey in a sea of blue — this is the backdrop to every scene as you cruise from island to island. The blue is so distinctive, even from a great distance, that American and Soviet astronauts, from their outer-space vantage points, have reported that the Adriatic is the bluest of the earth's seas.

Our tour, however, began in the north with a cluster of islands that guard the Kvarner Gulf: Cres, Lošinj, and their smaller neighbours: Susak, Unije and Sarkane. As our steamer churned seaward through a sultry morning haze, beneath us lay the mysteries of the underwater world. It was quite easy to imagine submerged ruins holding secrets of the past and, hidden in the murky green depths, the myriad forms of submarine life. The distant island, its contours gradually growing larger, presented another enigma. We were approaching Cres from the north, and its shores appeared wild and uninhabited.

With its steep cliffs dropping abruptly into the sea and spreading wooded hills, the island has a total land area of over 400 square kilometres. Our ferry docked at Porozina and we disembarked. On the road winding south the landscape took on a different aspect: less wild, greener, thick with clumps of broom, its yellow flowers gaily dotting the hillside in springtime. The fragrance of Mediterranean vegetation hung in the air. Here and there in the green scrub the white fleeces of grazing sheep reminded us that the warm hearth of a peasant cottage could not be too far away.

Pag

18. *From the Velebit channel the northeastern coast of the island of Pag looms rocky and inaccessible. Apart from lighthouses sited on headlands to warn ships of danger there is no other sign of life. The channel waters are navigable as long as the sea is calm but become treacherous when a fierce north wind blows.*

19. Above. *Hoping to hear the latest news, the townspeople of Pag like to gather on the main square in front of the parish church, a 15th-century triple-aisled basilica, the work of the well-known sculptor Juraj Dalmatinac.*

Below. - *A man and woman in the folk costume of Pag. The island women are well-known for the lacework which adorns their costumes and glossy black hair. During the long winter months their skillful fingers weave fine thread into intricate, fanciful patterns.*

20. *Pag is linked to the mainland by a bridge which spans the narrow Ljubačka straits to join the rocky headlands of Ošljak and Fortica. The bridge has given the island direct access via the Adriatic Highway to Zadar and Senj. Pag is the only large island in the Adriatic which can be reached from the mainland by a bridge. Regular ferryboat services run to the other islands.*

Inaccessible and isolated, Osor was once the seat of a bishopric. A gilt dome still crowns the cathedral. The town stands on a narrow isthmus that reaches out to the neighbouring island of Lošinj. A drawbridge spans the few yards to the opposite shore. When we arrived, Osor was basking in the sun, looking like a peaceful fishing village, but its abandoned houses and deserted stone piazzas told a different story — one of fire, pillage and earthquake. History and time had dealt harshly with the town. What remains stands as a challenge.

The magic remains of human endeavour.

In a silence disturbed only by the insistent whirr of the crickets, my thoughts turned towards the past, a very remote past, for men have lived here many centuries. The cave dwellings at nearby Punta Križa date from the Stone Age. But the island's first recorded settlers were the Liburnians, an Illyrian tribe whose stone fortifications and necropolises have yielded fascinating finds. Next to the fragment of a stone frieze a lizard was sunning itself, a miniature offshoot, so to speak, of some mammoth breed that inhabited these parts in prehistoric ages. At the time of the Greeks both islands, Cres and Lošinj, were known by one name — Aspyrtides. The Romans, however, gave this name to the main town, shortened to Asporos. As Roman traders unloaded their merchandise from the East, the town prospered and grew into an important centre of commerce and administration. When the Roman Empire was divided, Osor found that it owed allegiance to Byzantium. In its heyday, when it was important enough to have a bishop's palace, Osor boasted a population of eighteen thousand. Today it numbers eighty. The decline was gradual but inexorable. In the tenth century attacking Saracens set fire to the town, the flames gutting the houses and ravaging the splendid stone façades. The ruins we saw before us spoke eloquently of oppression and suffering: Venice and the tenth century, the Hungaro-Croatian king Koloman, pacts and wars, plague and malaria. Men had laboured in stone: temples, mosaics, villae rusticae, inscriptions, tombstones... A few more paces, and we walked out of the town. An old woman in black, stooping with age, murmured a greeting in passing, her voice so faint it was barely audible. By the cemetery, near a chapel with three

A contemporary and friend of Michelangelo, Bruegel and El Greco, in the service of Cosimo de Medici, Julije Klović — Clovio Croata (1498-1578) — left the Adriatic coastland at an early age for Italy where he lived and received his education. His works hang in museums throughout the world.

bells, were the exposed foundations of an early Christian cathedral — a twin-naved basilica and baptistery. Further on, close by the bridge, lay the ruins of a Benedictine abbey — a triple-naved Romanesque basilica built in the fifteenth century and dedicated to St. Peter. High above the cemetery dark, pointed cypresses swayed with the wind.

Instead of continuing on the main road to Lošinj, we headed for Punta Križa, the southernmost tip of Cres. Suddenly and dramatically, the cold stone ruins and shadowy piazzas were forgotten, and reaching Punta Križa we breathed deeply of the salt-laden sea air. Swinging inland the rocky shore forms a cove, its water reflecting a translucent green from the sea bed. I slipped on a breathing mask and without harpoon descended into another strange and silent world. A stroke of the arm, and the algae quivered. A surprised fish flashed its silver scales in fright, disappearing into the darkness below. Only a pair of eyes shone from the depths. Somehow it seemed to know that no danger threatened, that its pursuer would venture no deeper. Slowly the fish came back into the green light and swam towards a jagged rock under the surface of the water. It was a good-sized fish, a respectable haul, but I wasn't carrying a harpoon. The fish turned again and came towards me, the unwelcome stranger trespassing in its domain. I kept still, arms at my sides so as not to startle it. The fish drew closer and was almost within reach when, with a flip of the tail, it swung about. There it hung, its fins scarcely moving. It seemed to be asking what I was doing there. Our mute dialogue lasted only an instant. Needing air, I instinctively thrust upwards and surfaced. For some time afterwards I retained the memory of those eyes shining out of the obscurity. Another haunting souvenir was the dim outline of a sunken ship, covered with seaweed and barnacles. Bent, twisted iron, railings, gaping holes that once were cabins — a shipwreck is an ugly sight.

Once again we were following the shore past vineyards, olive groves, old stone houses, and in every yard a shady arbor of tangled leafy grapevines. In a tiny harbour with a small medieval church fishermen were drying their nets. A donkey brayed, persistent as a factory whistle. It must be time for lunch.

The shore twists and turns into so many coves and inlets that Cres has the third longest coastline of all the Adriatic islands. Numerous species of fish inhabit its waters. Punta Križa, Nerezina, Martinšćica, Cres town ... wherever people have made their homes, small harbours shelter rows of fishing boats, nets spread out on the waterfront drying until nightfall. Then by the light of twin lamps the fishermen cast their nets hoping to intercept schools of sardine, mackerel, or anchovy.

There is more, however, to the mystery of the island than deserted towns and crumbling ruins. In the centre of Cres, in sharp contrast to its parched and barren surroundings, is Lake Vransko Jezero, a huge reservoir five kilometres long and one and a half wide, over 200 million cubic metres of pure drinking water. But the lake has several peculiarities. First of all, the water stays fresh even though the lake bed lies 40 metres below sea level. The source of this water is not at all certain. It is generally believed that underground rivers from the mainland feed the lake and flow on under the sea bed emptying elsewhere. Another peculiar feature of the lake is that the water level remains steady during the heaviest rains and then, curiously enough, in periods of drought it rises. Yet some connection with the sea must exist, for eels are found in the lake and eels spawn only in the sea. The theory is that the pressure of the lake water exceeds that of the sea water and thus prevents the salt water from mixing with the fresh. Whatever the scientific explanation, Lake Vransko Jezero has inspired any number of tales of witches and spirits that haunt its shores by night, though the stories are dying out together with the old people who knew them. The young people of the island are too busy for story-telling: they are digging ditches to lay pipes for running water.

Many church treasuries on the islands contain valuable works executed with immense skill, talent and devotion. This is a detail of a processional cross in the Dominican monastery of Brač.

LOŠINJ
AN ANCIENT TRADITION OF SEA-FARING

Last of the western Kvarner islands, separated from Cres by a narrow channel, is picturesque Lošinj, its shores indented with dozens of coves and inlets. For thirty kilometres the road follows a slim strip to a cove which shelters the small coastal town of Lošinj where numerous rowboats, sailboats, and even powerful cabin cruisers are moored in the natural harbour which in its day was a busy, thriving port. Once again I turned to the past to learn more about the town and its former importance in the Adriatic.

Down by the waterfront, where virtually the whole town gathers after sundown, I came upon a group of old men, and among them two retired sailors. The others were either fishermen, shipbuilders, or artisans. Smoke was spiraling from the sea-captain's pipe as he puffed away, oblivious of the tourists clicking their cameras. These old people are a disappearing race, the last to tell tales of sailing on two-masters. Both the old sailors were over ninety. Their homes, they said, were real museums. The objects and new ways that the returning seamen brought home with them from their long voyages had a considerable effect on life in the prospering island communities. We entered the stone doorway of a house that had belonged to Anton Busanić who in 1843, flying an Austrian flag, had been the first captain from Veliki Lošinj to round the Cape of Good Hope. The navigator's portrait was hanging over a green marble fireplace. As evidence of his much-travelled life there were Florentine candlesticks, an ivory clock of Chinese origin and Japanese vases; large sea-shells from Indonesia, Hong Kong tapestries and pictures of Balinese dancers; small-scale models of sailing vessels, drawings and paintings of ships in stormy seas.

The sea provided a route to other lands, their cultures and their people. Sailors here have a saying: put your finger in the sea and you can touch the whole world. From childhood to old age, Lošinj seamen always maintained their contact with the sea. The sea was their livelihood and they loved it despite the hardships of a sailor's life. For island life the sea was of vital importance.

Lošinj's sea-faring tradition goes back as far as Roman times. The first fishing nets, however, were brought to

Lošinj by the Botterini and Ragusan families who had come from Croatia. They taught the islanders, whose principal industry until then had been raising livestock, how to fish. This occurred, according to official records, in 1640. Ten years later Captain Petar Petrina set sail for London and made a safe return, thus marking the beginning of Lošinj's maritime expansion.

Early in the nineteenth century the Napoleonic wars raging in Europe caused food shortages, which bolstered trade with Russia. Lošinj's ships transported wheat from ports on the Black Sea to England. Somewhat later they carried salt to Norway, and cod from Scandinavia to Mediterranean ports. In those days Lošinj shipowners deployed a fleet of 180 vessels, brigs and brigantines, the majority able to carry cargoes of over a hundred tons. The bristling maritime trade sparked another industry – shipbuilding. By 1870 six shipyards had built 45 vessels, two of them one-thousand tonners.

Those were times when men achieved sudden wealth, when Lošinj shipowners, builders, and captains were known as *paruni*, or big bosses. The old captains still talk about those days. Contemporaries of the period, about a hundred years ago, wrote that on their return the captains had new houses built of stone with intricate carving over the doorways and windows. They wore expensive clothing and silver-buttoned jackets. They commissioned paintings from fashionable artists and had their portraits done. Their children were sent to the mainland for a city education. In a word, they acquired the accoutrements and manners of aristocrats.

In the old part of Veliki Lošinj there still live a few seamen whose three-masters circled the globe. Mounting one of the steep, quiet streets I passed facades with tall Venetian windows partly concealed by the luxuriant vegetation. Oleanders bloomed in the shade of cypress trees, and filling the air was the mingled scent of rosemary and jasmine. Here and there sunrays pierced the twilight in spreading shafts of light. Here the sea-captain waited out his remaining days, his memory fading. If he recalled anything, it was the wind. When the *Medusa* sailed round the Cape of Good Hope, and later Cape Horn, the seamen knew that below the equator they would encounter first the Elysian and then the Levanter winds. Everything depended on the wind. And now the old seaman was waiting for a clement wind to bear him on a voyage of no return.

Countless are the tales of mariners and their craft. Our aging story-tellers are easily recognized by their navy-blue caps, nowadays sported mainly by yachtsmen. They sit near the chapel, or on the quay, their number continually diminishing. One sailor, just before he died, told of meeting a ship from Kotor Bay in Singapore one Christmas towards the end of the last century. And today they were telling their young listeners what a feat it was for a frail craft to sail round Africa all the way to Singapore. Ship captains had certain personal qualities and talents. They understood the vagaries of the elements; they could judge danger and make quick decisions; they had confidence in themselves and in the strength and precision of their hands, capable of performing the roughest physicals tasks and also of tracing the fine lines of a course on navigation charts, of handling a compass or a sextant. They were broadminded, calm and not afraid of hardship.

Historian Mijo Mirković attributes these qualities to the mountain folk who left their hills and settled by the sea, quickly adapting to new conditions of life. They were clever, skilled caravan guides, accustomed to the harsh natural elements that attend mountain living. The old sea-captains remembered the brave Austro-Hungarian naval officers — mostly from the mountain regions — of whom the best known was Janko Vuković, squadron leader in World War I and a native of Lika. And Maximilian Njegovan, the ablest admiral and commander of the entire Austrian fleet, who was also from Lika. Down through the centuries Yugoslav seamen have distinguished themselves time and again, and today they stand in high repute.

The opening of the Suez Canal and the coming of the steamship marked the beginning of the end. Sailing vessels have survived only as training ships for the navy; all that remains are the small, stout two-masters loaded with wine kegs and baskets of figs for the mainland, or sand for construction sites in the big ports. Sailing has become a sport, and the clipper ships museum models. The days of the old sea-captains are numbered; all they can remember are the winds and storms that etched their faces.

On a roughly hewn wooden table stood a bottle of red wine. Just beyond, a charcoal fire produced the tempting smell of fish *na gradele*, that is, grilled over crackling embers of grapevine, which gives the fish a special aroma. The smoke from the fire mingled with the smell of seaweed, of pitch from a pine tree on the shore, its needles almost dipping into the water. As we waited for our portion of fish, we tasted an Opolo, a heady Dalmatian wine from the island of Vis. The fishing village nearby consisted of no more than a cluster of houses with a breakwater to protect the boats from the waves. On the hillside above were the outlines of roofs and the tall bell-tower of Veli Lošinj. An owl hooted in measured cadence, and the sea reflected a string of dancing lights from the harbour.

After the decline of its fleet Lošinj's waterfront became a promenade not only for aging seamen but also for Austro-Hungarian ladies who came to breathe the fresh sea air. The island was proclaimed a health resort by the penultimate Austrian emperor Franz Joseph. His mistress spent her summers there, in a secluded villa which is still standing.

CRES

29. *On the island of Cres, the sun smiles down on the roofs of Cres town, a small trading port flanked by old houses of weathered stone and budding holiday homes in the verdant outskirts. Though men have lived on the island since prehistoric times, the first settlers of historical record were the Liburnians.*

30. *The elegant portal of a 15th-century church in Cres, one of numerous Venetian Gothic and Renaissance buildings in the town. In a web of narrow streets several churches shelter valuable works of art from this period, such as a 15th-century wooden* Pietà *and a polyptych by Alvise Vivarini.*

31. *Early morning in the narrow streets of venerable Cres. The tempting smell of baking bread fills the streets. The girl waiting has just had "first-hand" information that she'll soon get a fresh warm loaf.*

32. Above. *Twisting and turning into coves and inlets, the southern coast of Cres is covered with pine woods and fields thick with broom, its yellow flowers gaily dotting the hillsides in springtime. In the background lies the freshwater lake of Vrana.*

SUSAK
THE ADRIATIC'S ONLY ISLAND OF SAND

There was a heavy sea rolling when we left Privlaka Bay, and our launch trailed a wake of white foam. Behind us lay Mali Lošinj, its stone houses steeped in sun, coastal steamers and yachts moored at the quays, and in the shipyards the red hulls of skeleton ships. With hardly a whisper white-crested waves broke against the craggy headland.

The boat was pitching, its prow rising with one wave and breaking through the next. On the horizon I could already make out the grey hump which was Susak, our destination. I had visited the island some years ago and I still remembered the fierce north wind that was blowing the day I arrived. An hour's cruise to the island would be time enough to recall a few details about my earlier visit...

... The steamer cut through the waves and reached Susak

Below. - *Cres harbour, the liveliest spot in town, where most of the public buildings are located. In the centre are the main gate surmounted by a 16th-century clock tower and the town loggia of the same period, now serving as an open-air market for fruits and vegetables. Round the harbour are patrician palaces with handsome facades.*

VIKING
Marija

21
22
RIJEKA
PEPS

just before nightfall. There was a strong north wind, and the opal-green waves mounted higher and higher. The captain decided his ship couldn't dock in the shallow harbour, for the current might draw it onto hidden reefs. A boat would come out for me if the waves weren't too high. The ship stopped, and its searchlights picked out a fishing boat. The men in the boat were rowing towards a ladder which had been lowered over the side of the ship. I was the only passenger. It was already dark, and I heard the captain's voice: "Jump when the boat is on the crest of a wave." I jumped. As the lights of the departing ship grew smaller, I found myself in a raging sea with a pair of strangers pulling silently towards a speck of green light. I was soon on firm ground...

A four-square-kilometre island of sand, Susak remained isolated for many centuries. Its inhabitants developed, to an even greater degree than the other islands, a distinct dialect, customs, dress, and a personality of their own. The first time I arrived, night had already fallen; today we disembarked in broad daylight. The impression was the same: a strangely silent island somehow lacking in temperament. We met people and greeted them. They returned our greeting with indifference, showing no curiosity at all. They simply sat there, gaze fixed on the sea, or ambled off to the village, where the wind whistled through the reeds and the fences driven into the sand.

Susak is small; nevertheless, many books have been written about it. The Yugoslav Academy of Arts and Sciences in Zagreb has published a 1000-page treatise on the island. Year after year anthropologists and geographers came to take photographs and measurements, to make observations and collect data. The inhabitants welcomed them in their own fashion. When the researchers went to the elementary school to administer intelligence tests, they asked the children what a car was, what a train looked like. The children didn't know. Then a little boy stood up and asked the stranger what a *škeram* was. Of course, the professor didn't know that it is a small wooden lock to which oars are attached. The islanders enjoy telling this story, turning it to their own advantage and adding that a person isn't born to know everything.

In general the islanders give the impression of being self-

LOŠINJ

33. *Picturesque Lošinj, last of the western string of Kvarner islands, is separated from Cres by a narrow channel. At the end of a slim strip of land, surrounded by pine woods, lies Mali Lošinj, an important maritime centre in the Adriatic which the South Slavs have been sailing for centuries, as able in navigation as in battle.*

34. *At the foot of Veli Lošinj, in a quiet cove on the northern side of the island, is a small fishing harbour, just a few bright-coloured houses clustering round a stone breakwater where the fishermen moor their boats and spread their nets out to dry. As an occupation fishing attracts fewer and fewer people.*

35. Above. *Tiny fishing villages are becoming popular with holiday-makers. Friendships are made over a plate of grilled fish, at early evening song gatherings when church bells toll from ancient campaniles.*

Below. - *Family life is lived in courtyards, behind ivy-clad walls and old iron gates. When the sun sets and a refreshing coolness falls upon the courtyards, the scent of pines, rosemary and myrtle mingles in the air.*

36. *From an early age the islanders learn to climb steps. Clinging to hillsides, the island towns and villages are crisscrossed with stone staircases. An "up-and-down" way of life becomes natural, and islanders well advanced in years easily go up and down the hundreds of steps that separate them from the main square on the waterfront and the harbour where a boat awaits them.*

sufficient and uninterested in newcomers. One grasps and feels the full impact of isolation. Yet this exclusiveness applies only to people who come to the island, never when the islanders themselves venture abroad. The men of Susak have a long tradition as able seamen, especially at the helm. In Hoboken, New Jersey there is a street which, to all intents and purposes, might be another Susak, for ten times more people from Susak reside there than on the island itself. 280 people live here, among the terraced vineyards; in Hoboken there are 2,800! Furthermore, the American Susak has given generous assistance to the Susak in the Adriatic. Immediately after the last war, when the country was still recovering, the island received whole shiploads of goods including such improbable items as false teeth, which the islanders sold in the Lošinj marketplace, along with their figs and wine.

Susak is the only island in the Adriatic entirely composed of sand. A firm layer of limestone supports dunes which may rise to a height of one hundred metres. After the stony paths of the other islands it was like walking on a soft, thick carpet. But the question was: how was all that sand deposited in the rocky landscape we had grown accustomed to as we proceeded from island to island? It was almost as if a piece of the African coast had broken off and floated into the northern Adriatic. This geological mystery has puzzled many a specialist. Three theories have been proposed to explain the phenomenon; none has been proven. The island may have been formed by silt from the Po, Raša and Soča rivers. Or winds from the eastern Alps may have deposited sand particles on a rocky reef. The third theory is a combination of the first two.

On my first visit the cold north wind howled for ten days. There I was, cut off from the rest of the world. No one came to the island. The inhabitants prayed that the wind wouldn't destroy the vineyards. But they weren't content just to pray. They wove reeds into fences that would protect the vines from the wind. That winter salt water polluted the wells, making the water unfit to drink. The wine cellars opened their doors, and wine was served instead of water. Even food was cooked in it. On one occasion the sausages I ate for supper had simmered in wine.

Everything is affected by the sand. The women wear

lightweight slippers made of goatskin that resemble ballet slippers. On any other island shoes like these would not last more than a day or two. But here there are no stones. If you wanted to break a neighbour's window, you would have to throw a potato or some other object. The stone blocks for the houses were quarried on the nearby islands of Sarkane and Unije and transported by sailing vessel. The sparse vegetation consists solely of grapevines and reeds, without a trace of the usual Mediterranean growth: scrub, rosemary, wormwood; there are no olive trees, no fig trees. The grapevine, fortunately, thrives on sand, which accounts for the excellent wine that the islanders drink. They have more than they need but rarely offer any for sale.

There are many quarries of high-quality stone on the islands of the Adriatic, especially in the Korčula area, so it is not surprising that the islanders gained a reputation as stone-cutters. This relief showing figures on bended knees before the cross appears not only above church portals but also on the palaces of prominent patrician families.

Everything appears soft and light. There are no donkeys, no beasts of burden; the women carry everything on their heads. The reeds, however, serve the islanders well. They act as windbreaks; their pliant roots hold the shifting sand in place. They also make good fuel. Nonetheless, the island is blowing away. Every year, under the onslaught of wind and rain, Susak is several centimetres lower. Above the howling winter wind we could hear the rumbling of avalanches of sand. On these occasions the inhabitants say that Susak is sinking into the sea, not that they are really worried about it.

On the highest and safest spot on the island, which also commands the best view, the islanders placed their cemetery. Among the names carved on the tombstones there are few different surnames, only a dozen or so, four of them outnumbering the others. These four families, as tradition has it, came from the Crimea. One family, still quite large, is named Tarrabocchia, an italianized version of Karbog, or *karabogh*, which means "sick man" in an oriental language. By intermarrying, the island population managed to reproduce itself, and marriages outside that limited circle were rare and quite exceptional. The same combinations of genes, however, produced certain typical physical features: women with high, prominent foreheads and pale eyes.

That summer morning, after cruising for over an hour on a choppy sea, we finally set foot on land again. Susak was deserted, empty, inhospitable. We turned towards a house gaily painted bright red with a sign that spelled out in big letters *Broadway*. Its owner, one who had returned from Hoboken, serves good island wine and Coca-Cola. The tables under the red awning were empty. The owner's explanation was brief: it's Sunday, and everybody is in church. Religion forms part of their isolated tradion. In the old Baroque church I attended a very odd auction. After mass had been said, the congregation remained in the church while the priest auctioned off seats in the pews. "The second seat in the first pew is free. Josip Skrivanić has died, God rest his soul! I hear ten thousand... Do I hear twenty, thirty, forty... sold." Then the *Sansegoti,* as they call themselves, left the church. During the Italian occupation Susak was called Sansego. The inhabitants, whose dialect constitutes a mixture of Old Slavonic and Italian, still refer to themselves as *Sansegoti*. When I said I could not understand them, they just laughed. It is also their custom to assign a nickname to every stranger. I didn't learn what they called me during my ten-day winter visit until the steamer that I was to leave on had entered the harbour. *Očalin* is leaving, said my landlady. *Očalin* means "a man with glasses".

This is how an anonymous artist visualized a galley caught in a storm with the Madonna framed in clouds (17th-century Dubrovnik galley).

The Southern Kvarner Islands

KRK
A SUMMER'S
NIGHT SAIL

Upon leaving the northern group of the Kvarner islands one travels south with a certain feeling of expectancy. Constantly varying in size and shape, the islands succeed one another like a herd of whales migrating to warmer waters. Bare rocks and reefs, bleached in the sun, with an occasional patch of green scrub, lie scattered like breadcrumbs round loaves as large as mountains. These, too, are barren, though navigable channels and coves make them accessible. The main road along the coast winds in and out, and rounding each bend one has another surprise, a new view of the islands and archipelagoes bathing in the sea, each one more surprising than the last one.

Viewed from the mainland the islands might be ships at anchor guarding these coveted shores from hostile foes. The numerous coves and inlets protected by hidden sandbars, the headlands from which one could scan the sea, served as natural defences and allowed considerable strategical diversity as history has shown on repeated occasions. A thousand islands – a thousand strongholds. Besides the Slavs, who, having reached the Balkans, continued to the Adriatic occupying areas which the indigenous Illyrians had not settled, the coastland was fought over in turn by Avars, Greeks, Romans, Celts, Goths, Byzantines, Saracens, Turks, Venetians, Genoans, Austrians, French, Germans, and Italians, each destroying and then rebuilding in order to ensure their own presence and domination. The region exchanged hands again and again as countless battles were fought on land and sea. It is one of the ironies of fate that these sunny shores should have known war, epidemics, earthquakes and death on such a massive scale. The longest to maintain their domination were the Greeks and the Venetians, together with the Byzantine Empire. Yet the rule of the winged lion of St. Mark – figured on coats-of-arms with an open book which signified submission and slavery to every Slav — was not undisputed. The great Venetian fleet was harried by the Turks and by the corsairs, down to Napoleonic times.

Reaching Črišnjevo, where steep cliffs flank a narrow channel, we crossed on the ferry to Voz. In a few minutes we had an explanation for the fresh, man-made scars that were visible on the opposite shore, on the island of Krk, the largest in the Adriatic. Iron pilings jutted out from both the mainland and the island for a bridge which will span three kilometres of water. For if the past excites our interest, the future also holds much in store. Beneath the bridge there will run a pipeline pumping thousands of tons of oil over the rock-faced mountains to the interior. The airport with its roaring jets is already there on the island, and also a tangled mesh of silver pipes and towers — strange tokens of industrial development that still startle local shepherds and their flocks.

From this approach Krk looms so large that it no longer gives the appearance of an island, and we hastened across to the other side, to a familiar setting of ancient towns and pine-fringed shores.

When the ancient fortress walls of Krk town came into view, they appeared to be decorated for some fete. On closer examination it turned out to be laundry hung out to dry beneath the towers of the Croatian count Nikola Frankopan. Their resemblance to white flags of surrender was singularly inappropriate on this particular site, for in order to lure the count and his followers out of fortified Krk, the Venetian doge had been obliged to resort to trickery. The unsuspecting count boarded a waiting ship, never doubting the doge's word, and was carried straight to a Venetian prison and eventually executed. Now hanging from the battlements of the old Frankopan castle we see snow-white sheets swinging in the sea breeze. Beyond the walls is the cathedral, erected on the foundation of a three-aisled early Christian basilica and, going still deeper, the ruins of a Roman bath. Inside the cathedral one finds Gothic rib-vaulting, and, above the altar, a painting of *The Entombment* by an Italian artist, G. A. Pordenone. A painting from the seventeenth century, hanging on fifteenth-century walls, the whole resting on fifth-century foundations.

Towards evening a strengthening south wind was forcing the waves higher and higher. They thundered against the Frankopan walls where a group of fishermen's wives stood waiting, their gaze fixed on the turbulent waves. Windy gusts whipped their long black skirts around them as they stood there still as statues. A good part of their lives is spent waiting: they wave farewell to the departing boats and then wait for their return. But when the sea looses its fury, the wives climb onto the walls and stand there for hours — willing the fleet's return and their husband's preservation from death at sea.

The Frankopans were a well-known feudal family who came from the island of Krk where they had estates. The last member of the family was Franjo Krsto Frankopan (1643-1671) who was executed in Bečko Novo Mesto for having been involved in a conspiracy against the court. Having received a humanist education, he wrote poetry and was the first to translate Molière into the Croato-Serbian language.

Death at sea would be, of course, the best death for a sailor, or a fisherman. Those not claimed by the sea remain at their posts until their strength ebbs. Then they take their places on the stone benches, staring at the watery surface they knew so well. Their favourite topic of conversation is the weather. They are reliable barometers, their forecasts infallible. A breeze tells them if the weather is going to change, if the sea will be rough or calm. Yet when the women mount the walls, their faces strained with fear and uncertainty, the old fishermen seek shelter at the *osteria*. There they sit drinking their habitual mixture of wine and water, shaking their heads mutely until through the open window they hear a joyful cry, a sure sign that somewhere on the horizon the women have spied the masts of the returning boats. Then the old men push back their glasses and with unsteady gait walk to the harbour, waving their caps with pride. Come what may, their sons always make port safely.

The Frankopans dominated the island until the fifteenth century. There is evidence of their power and affluence everywhere, especially in the churches with their chapels and tombstones of illustrious bishops. The oldest is St. Donat's near Punat with a cupola dating from the eighth century. There is also Vrbnik on its eerie clifftop perch, or as the local folk call it "the big mountain," and then Malinska, Omišalj, and Baška, ancient coastal villages amidst green fringes of Mediterranean vegetation. One kilometre from Baška, its stone walls protecting a cluster of houses from the north wind, is Jurandvor with its pre-Romanesque Church of St. Lucia. The church was given to the Benedictines in 1100 by King Zvonimir of Croatia. To grace the royal endowment they commissioned Paolo Veneziano, a Venetian artist, to paint for the main altar a polyptych which is now kept in the treasury of the Academy of Arts and Sciences in Zagreb. Also placed in the treasury for safe keeping is a stone tablet bearing an inscription which states that King Zvonimir presented land to the Benedictine monks for a church dedicated to St. Lucia. This tablet, known as the *Baščanska ploča*, is the oldest legal record in the Glagolithic script.

Offshore, in the channels separating the islands, and even on the open sea, on calm, moonless nights, one can see clusters of moving lights. For a moment one may mistake them for some distant town or settlement. These are fishing boats which weave out of their sheltered coves just before nightfall, equipped with acetylene lamps that cast powerful beams onto the dark water. The lighted procession, repeated every night there is no moon, announces the beginning of another night's hunt for sardine and mackerel. The shouts of the fishermen and their wives, who that same afternoon in the scorching heat carried the flat fish-crates down to the boats, merge with the monotonous throb of the boat engines. And the fishing fleet departs, disappearing into the dusk, a many-beaded string of lights on the dark water.

This unusual procession has its origins in the Middle Ages. As man learned more about his natural environment he discovered that fish were attracted by light, and that a flaming torch would draw a whole school of fish to the shore. The next step was to use light to lure fish into nets. That was the beginning of fishing by lamplight, or *na sviću*, as it is practiced along the Adriatic coast. Originally, the lighted torches were carried in small boats which the men rowed slowly towards the shore where there were other, larger boats containing the nets. The operation lasted all night, and just before daybreak the nets were cast into the water and drawn towards the shore with the night's haul of fish. This practice continued until 1898 when a Dalmatian fisherman by the name of Kuljiš discovered that acetylene lamps could replace the torches. And that is the way it was the night we boarded a fishing trawler in the harbour of Krk town. It was a calm night, and old Tom expected a good haul. "It's just right, ideal for fishing," said the old fisherman. "There's no wind, and the water is warm and clear".

45. *The barren rocky wastes so common on the Adriatic islands are the result of erosion. Over the centuries the original forests were felled, and rains washed away the soil, depositing it in pits and hollows.*

46. *A hamlet with a population of 14. Its name — Košljun — is derived from the word* castellone *(little fort); its walls are still visible.*

Krk

47. Above. *Its cupola dating from the 8th century, the pre-Romanesque Church of St. Donat near Punta is one of the oldest on the island of Krk.*

Below. - *A typical landscape in the interior of the Adriatic islands. Clearing away pockets of soil, the peasants piled the stones onto walls to protect their minute plots from erosion and the wind. The grapevine thrives in these cramped conditions, and the tiny vineyards produce excellent wine.*

48. *Vrbnik, a small town on the northeastern coast of Krk, perches on a 50-metre-high limestone cliff overlooking the sea. Graves from ancient times have been discovered nearby. The town was fortified in the 14th century by the Frankopan counts, rulers of Krk.*

49. *The old town of Omišalj on the island of Krk is situated on a hill. During the Middle Ages it was fortified by the Frankopans who used it as a stronghold.*

50. Above. *At Jurandvor, about one kilometre from Baška on the island of Krk, is the early Romanesque Church of St. Lucia, built by Benedictine monks in 1100 on land granted by King Zvonimir of Croatia. A stone tablet known as the Baščanska ploča, the oldest Croatian record in the Glagolithic script, was found here.*

Below. - *The small town of Baška on the southeastern coast of Krk has 860 inhabitants whose principal occupation is fishing. The remains of Roman settlements have been found nearby.*

51. *The harbour of Baška opens onto a channel swept by fierce winds from the Velebit mountains. Long ago its inhabitants built a sturdy breakwater where their fishing boats could find safe haven.*

RAB

52. *On the island of Rab, last in the northern group of Adriatic islands, is a town bearing the same name, its beautiful setting dominated by the spires of medieval churches. The tallest of the four bell-towers which were erected between the 11th and the 14th centuries, belongs to the 12th-13th-century cathedral. The campanile rises in series of Romanesque twin-light, triple-light, and four-light windows ending with a balustrade and a pyramidal spire.*

53. Above. *Nearby is the triple-naved Romanesque basilica of St. Mary's, one of the most beautiful in the Adriatic. It was consecrated by Pope Alexander III in 1177.*

Below. - *A view from the cathedral bell-tower of the roofs of the town and harbour. It is a memorable experience to climb one of the campaniles when the bells toll and the startled pigeons sweep off into the azure sky, settling on the drains of the old roofs and balustrades.*

54. *When the sun sinks behind the horizon and the twilight tinges the church spires with gold, Rab turns an ear to sounds of the night: an owl hooting in the city park, the beat of drums from night spots.*

55. Above. *The costumes of Rab differ from those worn on other islands in certain details. A common denominator is seen in the black-and-white contrasts and the embroidered caps and aprons of the women's dress.*

Below. - *One of the three main streets in the old quarter. The stone escutcheons over the arched portals and balconies represent patrician families and artisans, each indicating his craft with a tool.*

56. *As one of the oldest tourist attractions in the Adriatic, Rab has long been singled out as a romantic spot where sentimental experiences meet with the changing times in a beautiful Mediterranean setting.*

The small fishing boats with their twin lamps were being towed out to sea by the trawler. I got into the nearest one, trailing my hand in the water. There was a sudden spray of luminous plankton, millions of particles of life dispersed in silvery cascades. The voices of the fishermen were muffled, unintelligible above the sound of the engines and rushing water. We sailed on, passing the isle of Ilovik. I knew that shortly before midnight, somewhere near the island of Rab, the engines would fall silent, and the fishing boats would spread out in a circle of light to begin their hypnotic dance with the fish. I returned to the trawler where young Ivančić, whose father and grandfather before him had been fishermen, stood drowsing at the wheel. "We were poor folk," he said, "like all fishermen." Life was a hard and cold struggle with the sea, and when the fish had been taken to market, little remained for their families. The fishermen of Krk still sing songs of bitter comment on the hardships of a fisherman's life. From a paper packet Ivančić produced a piece of dried fish. Old Tom used his penknife to open a bottle of red wine. The meagre repast was placed on a wooden ledge by the wheel. It was two in the morning, and the sound of the engines died. Captain Stipe directed the small boats into position through a foghorn. They called him Sven because he had spent many years fishing and salting cod in Scandinavian waters and had swollen nailless fingers to show for it.

The lights of the acetylene lamps danced on the calm surface of the water. In the centre of the lighted ring was the silhouette of a larger boat known as a *leut*, filled to the gunwhales with nets. Peering into the light the men tried to see if the fish were rising. Whole schools of fish will rise from the depths towards a light, but it is a process that takes hours. The only sound was Sven's deep baritone asking impatiently, "Are they rising?" But there was no answer. Under Sven's direction we jumped into a dinghy and rowed towards one of the now distant lights, speaking all the while in whispers for fear of scaring off the fish. It was an unforgettable sight. Like the plankton earlier that evening, the glittering scales of thousands of fish shimmered in the water. They drew closer to the light, like people drawn from their houses at the first warm rays of the sun. "There will be enough to fill three freight cars," Sven estimated with satisfaction. The fish lay three metres deep, and it was the same under each of the nine boats. If the fish did not rise high enough the nets would not surround them, and it would be a poor haul. Furthermore, summer nights are short and the first rays of light disperse the fish. A tardy decision could cancel a whole night's labour. The fish were in no hurry, approaching the light slowly. Old Tom was puffing nervously on his pipe. The fishermen's calls from their boats became more frequent. Hesitating no longer, Sven called out "Cast the nets!" and the men were roused from their lethargy. The *leut* began dropping the huge nets into the water. The trawler started its engines, the pulley ropes tightened and strained under the load of the full nets.

All at once we heard the cries of seagulls. "The first sign of daybreak," said Sven. The nets must be dropped before the seagulls begin circling overhead, their wings pale blue and the sea metal-grey. The first light of dawn could be seen in the east. No one paid any more attention to the small boats which had completed their task of hypnosis. It was now the turn of the pulleys and the hoists. The men in their yellow slickers and high rubber boots tugged at the ropes drawing in the nets.

It was morning and the sun was warming the rocky peaks of Velebit on the mainland. Knee-high in fish, the men packed crate after crate. When the trawler docked alongside the stone quay of Krk, the women were waiting to transport the crates in time-honoured island fashion — on their heads — to the fish market. Pleased with his haul, Sven took time for a cigarette. Together with the other fishermen, Ivančić washed down the decks, and old Tom was thankful for one more night's work.

By the time the sun was high in the sky, the fishermen had left for a few hours of well-earned rest. And on another island the sun was also shining — on the island of Rab, the next chapter in our story of the Adriatic islands.

RAB
Four spires in the sun

How can one describe Rab? I remember earlier visits, the enchantment of the ancient town, the grace of its portals and stone staircases. Caught briefly in the sunlight, Bacchus gazed down from a coat-of-arms carved on a nobleman's palace. For a few short minutes in the late afternoon the sun's rays pierced the shadows of the narrow street and fell on the stone doorway. A reddish tinge spread over the face of the wine-loving god, and for an instant its stone cheeks flamed like the huge crimson ball that was hovering over one of Rab's four church spires before dropping behind the pine forests. Suddenly I knew how to describe Rab — four spires in the sun.

Once again we delve into the past. Stone facades on either side of the street bear coats-of-arms, each with its symbolic representation of family history, each unique and handsome. These crests of patrician families are found all along the main street, one of three running the length of the town. Connecting the three streets are sloping passageways intersecting at right angles. Within the ring of walls town life is conducted beneath archways and on stone balconies, in the shade of a fig tree or under a boat canvas. Everywhere present is the romance of a bygone age.

Armorial crests dress many patrician palaces on the islands of the Adriatic. This escutcheon composed of a stylized crown, sword and cross is on the island of Brač.

Every Sunday the town Loggia, an open portico with six Renaissance capitals built in 1509, serves as a meeting place where the old people continue a centuries-old tradition: it was on this spot that the town notables reviewed all offenses, condemning the guilty party and exonerating the innocent. Once a week they pass through the old Sea Gate, near the twelfth-century Prince's Palace, to take their places on the stone bench of the Loggia, prepared to hear, or tell another true story about Rab's eventful past. Next to the Loggia is the Gothic Church of St. Nicholas, patron saint of mariners, protector of those who appeal for his aid in storm-whipped seas. Nowadays there are just as many storms, but the ships are larger and can weather them more easily. The old seamen, however, are fewer in number, as are the story-tellers.

From the Loggia a steep street leads to the oldest part of the town, Kaldanac, where parts of the old walls are still visible. These crumbling ruins mark the end of one era and the onset of another. In the fifteenth century Rab boasted palaces for the bishop, nobility and wealthy merchants, palaces bearing the crests of the Nimiri, Cassio, and Galzigna families. On the Domus Nimira facade is a Roman relief of the two-faced god Janus, one face turned seawards and the other towards the heights of Velebit where the Slavs had come from in search of a more rewarding life. In the mid-fifteenth century the plague struck, and Rab's civil and episcopal brilliance came to an end. The town walls were deemed high enough and the gates securely locked, but the dying town was never to recover. Spiders and snakes moved into Renaissance palaces and Gothic churches. Wherever the sun penetrated, ivy climbed over the ruins. Kaldanac was doomed and deserted. Outside the walls men set to work building another Rab. The coats-of-arms, however, were more likely to represent traders and artisans than noblemen and bishops. The town grew and expanded like a stone flower with churches and convents, the homes of the townspeople and workshops of artisans.

Isaac Newton and Rudjer Bošković had a high opinion of the work of Mark Anthony Dominis, bishop and mathematician, who was born on the island of Rab in 1560. Because of his religious beliefs he was declared a heretic, exiled and imprisoned. He died in a Roman prison on December 12, 1624, and his body was subsequently burned in public.

The Napoleonic wars put an end to its expansion, and in the early part of this century Rab welcomed the first tourists, making it the oldest resort in the Adriatic. Sun-bronzed travellers arrived by the boatload, and people who once looked to the sea for a livelihood found the tourist trade more remunerative.

Moored at the waterfront, amidst all the yachts, are some small white rowboats, each with a colourful canvas canopy as a sunshade. There the old boatmen sit, not even bothering to offer their services with a "Bitte schön, Barke fahren...." I stepped down into one of the boats, and its owner looked up, hoping he had a fare. He put aside his bowl of fried fish, his wine, and wiping his mouth with a handkerchief he politely indicated a red cushion with embroidery worked by skilled, diligent fingers. But what I wanted instead of a boat ride was to talk about the life of the boatmen, once so typical of Rab.

Like the white-gloved coachmen who used to wait before the Vienna opera house and at the final notes of Lehar's *Merry Widow* urged their white horses to the main entrance, the boatmen of Rab, decked out in straw hats, striped shirts and loosely tied red scarves, won the admiration of their passengers. "Bitte schön, Barke fahren" was an invitation to romance. But Opels and Mercedes have long since replaced carriages in front of the Vienna opera, and the white canopied boats are giving way to the high-powered, unromantic engines of Johnson, Evinrude and Tomos. As the boatman quietly finished his fish, a passing speedboat set the old boat rocking. "Bitte schön, Barke fahren," he said more out of habit than hope.

Here and there in the harbour sailboats float on the water like resting seagulls. In the town beneath the walls of ancient palaces the fishermen's wives sit watching the tourists as they take pictures of the stone coats-of-arms. In no way connected with the Morosini counts or the Nimiri family, the women had never wasted any love on the lion of St. Mark. Yet they love their stone city which now, after the daytime bustle, was preparing for rest.

From the green pine woods and hotel terraces came the sound of pop tunes. Youth has overcome tradition, and the summers no longer belong to the boatmen. Relaxed young bodies in faded jeans sway to rhythms that summon up the heart of black Africa. Water-skiing or speedboating by day, many choose nudism so as to expose their bodies fully to the sun. Free of the constraints of clothing, they seek closer communion with nature, the warm touch of sun-baked rocks and the sea.

With palaces and escutcheons, churches and spires, with its fifteenth-century Church of St. Euphemia and convent of cloistered nuns, its aging boatmen and naturist sun-worshippers, Rab lives in the present yet surrounded by the past.

Under the cover of darkness a night owl hooted in counterpoint to the drumbeat of the dance bands, and on the morrow islanders and visitors would again hasten down to the sea.

Four spires in the sun.

PAG

LACE FROM PAG — A SKILL LEARNED LONG AGO

In a checkerboard of half-submerged flats bordering the sea — a picture vaguely reminiscent of the rice paddies of Indochina — lie the salt pans of Pag. Their harvest is gathered in the driest months of the year when there is little chance of rain, which would be fatal for the crop. This contest with nature goes far back in time, for the Romans extracted salt by exactly the same methods employed today. The sea water trapped in the salt pans evaporates in the sun, leaving the salt which is then raked into white mounds. It is unbearably hot work and it is not always easy to outwit nature. Loaded onto carts, and more recently onto lorries, the salt is taken to Pag town where it is ground, packed and transported to the mainland — every year over eighteen thousand tons of salt.

Approaching the island across the channel from

Karlobag was an awe-inspiring experience. The barren grey rock showed not a trace of life. Making a sharp turn, our steamer rounded a craggy headland and entered the narrows that protect Pag bay. Dominating the promontory of Zaglav were the ruins of Roman fortifications from the first century A.D. Built to ward off Liburnian attacks, they still keep watch at the entrance to the bay.

In medieval times the towns of Zadar and Rab both claimed the island of Pag and tried to settle the dispute by force. In 1071 King Krešimir IV of Croatia divided the island and accorded one part to Zadar and the other to Rab, a division which prompted further contention.

This decorative 17th-century map of Pag shows a battle taking place on the sea and also the low narrow part of the island where the salt flats were already being worked.

The church bell struck three that scorching summer afternoon. The lengthening shadows were just touching the rose-window above the church portal. It was around this church that the new town was planned in 1443. Its main street forms a perfectly straight axis of stone, as narrow and straight as the streets intersecting it. Apart from the church, which combines elements of Romanesque, Gothic and Renaissance styles, one also finds in the main square the Prince's Palace, with slender Venetian columns and arcades.

Two women in folk costume stood talking beneath the rose-window. Their black costumes were elegant and restrained, with lacework on the caps and bodices. Behind them in the shade of oleanders was a bronze statue of a stone-cutter, one hand swinging a hammer and the other holding a chisel. It was he who was responsible for the stone figures of saints and the portraits of fishermen, peasants and their wives. Juraj Dalmatinac, a well-known fifteenth-century sculptor, also planned the layout for the new town. The church, too, is his work, the stone tracery of the rose-window an inspiration for the hundred-odd women who make Pag's famous lace.

Seated on stools in front of their homes, the women work for hours over their round velvet cushions. The lacey motifs of the rose-window appear beneath skillful fingers swiftly drawing fine thread into intricate patterns. Engrossed in their handicraft, the women work in silence, with only a word or two to a neighbour. This skill was learned long ago to occupy idle winter months when there was no field work to take the women out of their homes.

Flocks of sheep wander over the hillsides and browse on low bushes flattened by the wind. The scrub on which they feed is salty: spray blown from the sea settles on the bushes and salts them. From the ewes' milk the women make a sharp, highly prized cheese, its quality due as much to its traditional manufacturing as to the salt pastures of this island with sparse vegetation.

In a parting image, Pag stands out barren and bleached against the blue background of the sea, the wind shrieking while the women quietly and methodically twist and knot their threads into fanciful lacework, adornment for their somber costumes and glossy hair.

The image of a galley carved on a tombstone on the island of Silba (1679).

65. *A lonely pair on the blue horizon. These two islands of the Central Dalmatian archipelago, from the air, look like two loaves of bread. The latticed panorama illustrates well man's heroic struggle to preserve the soil and its produce. Stone walls portion out the meager soil into hundreds of tiny plots which support peaches, almonds and figs, a few olive trees and medicinal herbs.*

KORNATI ISLANDS

66. *The Kornati archipelago — hundreds of islands strewn over the dark blue sea, like white clouds dispersed by the wind. Seen from the air it is an unearthly, lunar panorama. The archipelago consists of 8 islands, 109 islets and 30 reefs.*

67. Above. *The Kornati isles are not inhabited all year round. About 250 stone cottages tucked snugly inside coves on the larger islands are used to store fishing equipment and shelter sheep.*

Below. - *The soaring cliffs of the island of Lavsa were chiseled into their bizarre formations by the abrasive action of the waves on limestone rock. It is a dramatic, never-ceasing combat between the cliffs and the thundering waves.*

68-69. *The maze of channels in the archipelago is full of fishing boats winter and summer. Numerous species of fish abound in the Kornati waters. Ruins on the islands and in the shallow channel off Piškera island indicate that man has been fishing in these waters since Roman times.*

70. Above. *Spreading across an isthmus on the island of Piškera are the walls of a Roman toll-station. Soldiers guarded this fort which controlled the approach to the islands; fishing rights were paid in kind.*

Below. - *The homestead of Ivo Turčinov, or Pikolo for short, a one-story stone house on the island of Smokvica where the fisherman spends the summers with his wife Golubica. They leave when the autumn squalls bring the first rains to this barren, arid island.*

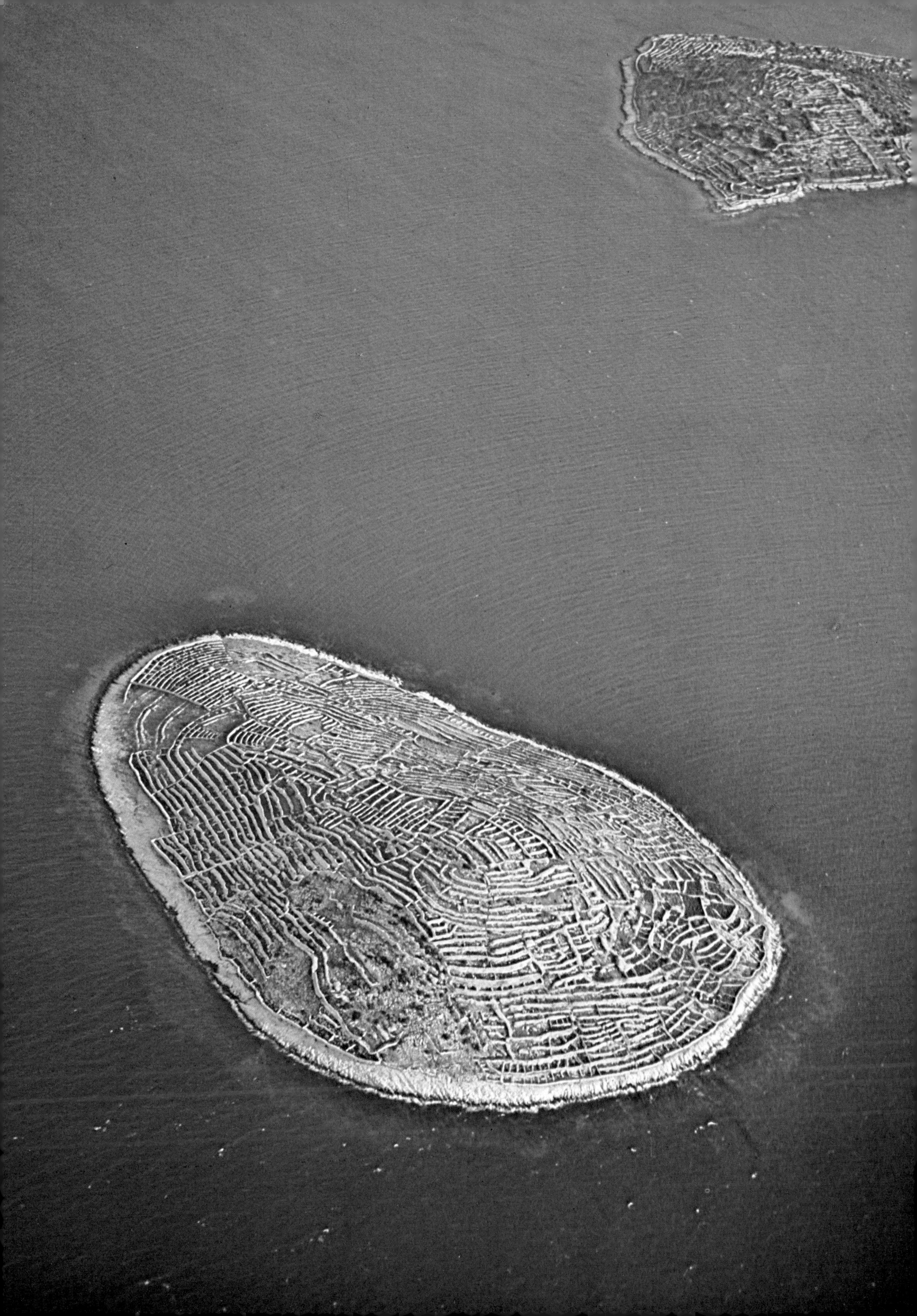

The Central Dalmatian Archipelago

KRAPANJ AND ZLARIN
Islands of coral- and sponge-divers

Coral beds are not plentiful in the Adriatic. For a long time their location was kept secret by the inhabitants of two minute islands, Zlarin and Krapanj, unique in the Adriatic. Fanning out from the entrance to the bay of Šibenik are the islands of Prvić, Tijan, Lupac, Zmajan, Sestrice and Zlarin. Further south, practically hugging the mainland coast, is Krapanj. It is not known exactly how the coral- and sponge-divers began their unusual trade. Some fisherman from Krapanj must have found an underwater bed of coral and, knowing its value, brought the first samples to the village. At any rate for a good hundred years the men of Krapanj have been combing the sea bed, picking up this difficult skill. Looking for coral, they came across sponges, a discovery which determined the future of both islands. The men opted for the perilous task of diving; the women took over all the other chores of island life. Their paths separated. On summer days in the villages of Zlarin and Krapanj one meets only women. If you ask them where the men are, they will lift a hand and answer: "Somewhere out there on the sea." "Somewhere" is among the hundreds of Kornati and Pakleni islands stretching all the way to Mljet. Like latter-day Ulysses the men follow coral trails, not returning until autumn. In springtime when the men prepare to leave, the women are absent. They accompany the sheep across the channel to pastures above the village of Jadrtovac and there, in the sheepfolds, make cheese from the ewes' milk. After their husbands' departure the women work long, hard hours, cultivating tiny pockets of soil, spinning wool and waiting until homecoming day in the autumn. Year after year, for an entire lifetime, nothing ever changes in the long-established routine imposed by an arduous struggle for survival. The voices of women calling to each other echo from the walled

Kornati islands

71. *An aerial view of the cove on Smokvica with the jetty and homestead. The large concrete platform serves to collect rainwater. In these dry regions it is the only way to collect enough water in the cisterns during the autumn and winter rains to last through the hot summer. Besides fishing, people on the island also cultivate olive groves, vineyards and fruit.*

72. *The low straight walls mark off property limits. The stones for the walls and terraces which support the soil are all gathered by hand. Despite the lack of water, fruit and medicinal herbs do well on the Kornati isles.*

73. *Sailing among these white isles and soaking up the sun on these rocky shores offer unexpected pleasures. It is also an opportunity to get away from the civilized amenities, hotels and crowds of the mainland and live freely in nature's environment.*

74. *The inside of an old fisherman's house on the island of Jadra: a typical rustic interior with an open hearth, shelves for dishes, wooden benches and a table. Accustomed to his lonely life, the old fisherman is one of the few who do not leave the island even in winter.* →

terraces that overlook the shore. Pausing for rest, they let their glance fall on the boats moored at the rocky shore. Their swinging hoes strike dull, hollow blows and sparks fly. The women curse under their breath: there are more stones than soil in the tiny plots. Some have been abandoned, overgrown with grass; sheep graze here, alongside goats who nibble at the low shrubs. The women spend their lives going back and forth from the island to the mainland, standing upright in the boats that bear donkeys, sheep and goats across the channel. A strong pull on the long oars is needed to propel the heavy, blunt-nosed boats through the water. The women guide them skilfully and when they tie up at some breakwater, one thinks of Noah's ark. The animals are seaworthy, capable of riding out storms, but they are not overlarge. They set out in groups of a dozen or so, supplied made secure in folds and stalls on the island, it is a sign that the autumn winds and storms are approaching and that soon the long summer's separation will be over.

Summer on Krapanj and Zlarin is picturesque. The women wear costumes consisting of wide, pleated black skirts and white blouses with yellow piping. The scarves binding their long hair are also yellow. The women are not at all eager to talk about their husbands and their work. When I asked, they looked at me in surprise wondering why I was interested in people who, after all, were only doing their jobs.

With the first summer days the divers of Zlarin and Krapanj prepare their boats for a long expedition. The boats are seaworthy, capable of riding out storms, but they are not overlarge. They set out in groups of a dozen or so, supplied with enough water and food to last some time. In earlier days they carried barrels of oil which was cast on heavy seas in the hope of calming the waves, just as the Romans and Phoenecians used to do. Wine and dried meat, together with fish, form the mainstay of their diet for the long summer months. The divers employ ancient methods, still using the *inželj*, a medieval apparatus for raising the coral from the seabed. Two wooden poles are crossed and firmly fastened at the centre, a long rope fringe attached to the ends. When a coral reef is discovered, the *inželj* weighted with a big stone is lowered into the water and dragged along the sea

KORNATI ISLANDS

75. Above, left. *In each of the coves where fishermen set up housekeeping during the summer months one sees oddly shaped wicker baskets. From the large openings at the top they narrow towards the bottom. Their purpose is to trap lobster.*

Above, right. *Cleaned and stretched out to dry on posts, squid are a tasty Dalmatian speciality. The squid are dried for use in the winter months when they are less easily caught.*

Below, left. *A daily task which consumes a good deal of a fisherman's time is getting his nets ready for the next night's catch. After cleaning them and letting them dry in the sun, the fishermen have to examine the nets carefully and mend even the smallest holes using a special needle.*

Below, right. *Fish are caught in many ways. The Adriatic fishermen adapt the technique to the season and kind of fish they hope to catch. One method is trailing a* panula *fixed with many hooks behind a moving boat.*

76. *A street in a fishing village on the island of Murter. Generally one storey or at the most two, the houses each have a terrace and wine-cellar in front. For decoration an old stone press for making oil is propped against the wall.*

bottom. The sharp coral points are caught in the rope which is then hauled to the surface. Often, however, the *inželj* catches on a snag and divers must go down and bring up the coral themselves. The coral beds, so rare in the Adriatic, are well known to the Zlarin divers. Their location was a well-kept secret passed on from father to son. When the boats headed for one of the beds, the men kept a lookout, and if another boat was following them, they would immediately change course and lose their pursuers. But all this belongs to the past. There are no more secrets, nor is anyone else interested in the dangerous occupation of diving for coral. Few divers live to enjoy old age. The occupational hazards are pressure sickness and heart trouble. They sometimes have to go down thirty metres in order to find the coral or free a snagged *inželj*. Danger and surprise also lurk in the shadowy deep. While a pair of divers searches for coral, their companions overhead keep their eyes on the sea. If shark fins are spotted, a few jerks on the cord warn the divers of the approaching danger.

The Tintoretto Madonna and Child with Saints *no longer hangs over the altar of the Dominican monastery near Bol on the island of Brač; it has been placed in the museum. This is a detail of the painting.*

Even today the old apparatus for raising coral has not been discarded for some more modern invention. Harpoons can be employed on shallow reefs, but the *inželj* is still in use. Near coral reefs there are often sponges which are just as difficult to raise from the bottom as coral. Divers have to search the seabed since they are hard to see from a boat. The sponges are cut away with knives, and even hammers are used to split the rocks.

When the sails billow with the first autumn winds, the flotilla returns, and the women breathe a sigh of relief, glad that another season is over and the men are safely home. The sea taught these men how to dive and today they travel far and wide undertaking difficult underwater operations. Many have worked on raising shipwrecks or bringing up objects from sunken galleys. At home they usually have a few amphorae as souvenirs of some scientific expedition. The Tanfara family of Zlarin was well-known in Mediterranean diving circles for its expert divers. The last of this diving family died ten years ago. It is an occupation fraught with danger, and in the cypress-shaded Zlarin graveyard lie many who died before their time, in the prime of life.

We encountered no divers in Krapanj. All were at sea.

The sponge factory on the waterfront was closed; its doors would open in autumn when the divers made port. The factory belongs to the local sponge- and coral-diving cooperative. In one of the houses a young girl was polishing coral. From the largest, many-spiked pieces to the smallest points, they are filed and polished, prepared for sale. Necklaces, brooches and curiously-shaped ornamental pieces are sent all over Europe, displayed in the jewelry shops of Venice and Vienna, Paris and Rome, where they are sold as Adriatic coral. The names of the divers and the islands of Zlarin and Krapanj remain unknown.

THE KORNATI ISLES

Drama in stone

We took off in a small, two-engine plane from Split airport. It was a sunny day, windy and clear, not a cloud in the sky. Flying over Šolta our pilot pulled up on the throttle and we glided almost noiselessly over a many-beaded archipelago, hundreds of islands strewn over the dark blue sea. Seen from the air, the Kornati islands are an unforgettable sight, an unearthly, lunar panorama inviting us to flee the cares of the world and embark on a Robinson Crusoe adventure. There are as many islands as there are days in a year, say the local inhabitants. This is speaking figuratively, of course, though the estimate is not too far off the mark. Map-makers have never been able to agree on the exact number of islands, each drawing the limits of the archipelago differently. According to old surveyors' maps the Kornati archipelago consists of 8 islands, 109 islets, and 30 reefs, making a total of 147, the largest group of islands in the Adriatic, and one would be hard put to find as many islands so closely packed anywhere in the Mediterranean.

Our plane dropped low over the islands. We tried to pick out the ones we knew, seen for the first time from the air. They passed beneath us quickly, dizzily. There was Piškera, half verdant, half barren, a few fishermen's cottages tucked inside a cove. Forming an outer seaward string were Gustac, Vela and Mala Panitula, Klobučar, Vodenjak, and many others. A little beyond rose Lavsa, round and more substantial than the rest. Just off Smokvica a boat lay at anchor between two slim tongues of stone.

The Kornati isles are not inhabited all year round. During the summer months people from neighbouring islands come to work the fields, put sheep out to pasture, or gather olives. There are some 250 stone houses on the islands which come alive in spring only to be deserted again with the first winds of the autumn. The scathing winds leave little vegetation on the islands, their bare rocky formations attacked unremittingly by waves and currents.

Buffeted by the wind, our little plane gained altitude, and we had a magnificent view of the whole archipelago – islets the size of white caps dotting the sea, the larger ones resembling giant stone lizards stretched out in the sun. It was a scene of extraordinary natural beauty tempting us, urging us to explore at close quarters what we had seen from our plane high up in the air.

The island scene was just as impressive at sea level when a few days later we arrived on Murter. Many of the tiny Kornati isles are owned by the inhabitants of Murter whose principal occupations are fishing and building boats. Our first evening was spent with Šime Matulica and his wife on their arbored terrace. There was a starry sky overhead – and loops of dried garlic and onions. A donkey brayed raucously in its stall below. Barn smells mingled with smoke from a fire where our supper was being grilled. Šime was explaining how to prepare fish. The quickest way was to fry it. But to bring out its full flavor one should make a *brodet,* a kind of fish stew. The recipes for fish stew vary up and down the coast, but the taste is more or less the same. Crabs abound in the Kornati waters: cooked *na buzaru*, that is boiled in salt water, it is hard to decide which is tastier, the crab meat or the broth in which it has been cooked. And there is nothing quite like wine from one's own vineyard!

Šime put it very simply: "The wine barrel is in the cellar. You can pick your own grapes. And the fig tree isn't too far off. Help yourself."

Help yourself – words of welcome that are often heard on the island of Murter. There is wine in abundance, and a few glasses set Šime talking. Thick fists resting on his knees, he spoke with gravity of the old days. He told us that the coastland was sinking at a rate of one metre every thousand years, that sea water had flooded valleys and depressions, split larger islands into smaller ones. At one time Kornat, Katina and Dugi were one island extending seventy-six kilometres, and entire settlements now lay under water. Sime turned the fish on the grill and continued. "You'll see," he said, "when you go out to Mala Proversa, a Roman jetty completely covered by water, and at Piškera, house foundations at the bottom of the sea. We'll all be under water one of these days." I asked the old fisherman how the archipelago had got its name. He paused for a second. It was named after the largest island, Kornat. Since the islands were inhabited only for short periods by people from other islands (Iž, Pašman, Dugi, Zlarin, Žirje), they all had different names for the archipelago. On Venetian maps the islands were marked as exceptionally dangerous for navigation. Šime went into the house and brought out a book. By firelight we read that the islands had first been mentioned by Constantine Porphyrogenitus in 950. He called them Grebeno. Its present name, however, is derived from the Italian word *corona* meaning crown; namely, the islands are *incoronati*, or crowned.

Šime brushed olive oil over a good-sized red mullet and placed it on a tin platter. We ate in true fishermen's style: the host split the fish along the backbone and with bare fingers everyone helped themselves. Greasy fingerprints marked the wine glasses which we raised toasting the health of the old fisherman and his wife, and the prospect of the next day's boat trip.

Milan had a good boat. Šime had already set aside fish for the next day, plenty of wine, cheese, ham, brandy, all that was necessary. We would be away all day. Milan had fished in these waters for years and knew them well.

We left just before sunrise, the hoar-white islands in front of us tinged with red. A seagull's cry announced that it had found a fresh morsel for breakfast. The whole scene was one of strange, unreal beauty.

Milan, a hefty fisherman in his fifties, toasted the new day with a swallow of brandy. Spread out on the deck were slices of cheese, as salty as the sea water which the sheep drink. Sheep subsist on the sparse island growth, and when the waterholes dry up they turn to sea water. Their milk is therefore heavily salted as is the cheese made from it. At sunrise we were already near Opat Point, and with a turn of the wheel we rounded the promontory and approached the island of Smokvica. To the left were Žakan and Prduša; seawards rose the hulks of Vela Kurba (Streetwalker), Babina Guzica (Grandma's Backside), Volić (Ox). I asked Milan how the islands had acquired such indelicate names. He shrugged his shoulders and said that some people had an odd sense of humour. The islands were bare and arid, awesome in their isolation. Long ago the Kornati isles were covered with forests. That was not within Milan's lifetime but he had heard of a great forest fire in the middle of the last century which had burned for forty days. Borne by the wind the sparks leapt from island to island so that from the mainland it looked as though the sea itself was on fire. Vegetation never took root again.

The ancient Greeks and Romans who settled on the Adriatic islands brought with them their own cultures, as numerous archeological finds have well established. This bronze head of a goddess found on the island of Vis shows clearly the influence which Greek sculpture had on Roman works.

We dropped anchor in a cove: two stone houses, a breakwater, and nets drying in the sun. Climbing the slopes were terraced vineyards, olive groves, almond and peach trees. The verdant northern part of the island contrasted sharply with the rock-covered southern half. The hillside above the cove commanded a splendid view of the surrounding islands. This green oasis, an example of man's need to mold nature to his own intent, was the homestead of Ivo Turčinov, or Pikolo for short. He was mending nets on the shore, and his wife Ana — her nickname was Golubica — held up a huge lobster trapped only that morning. The archipelago waters are excellent fishing grounds with plenty of lobster and crab, highly sought after for they fetch a good price at the market. In each of the coves where fishermen set up housekeeping during the summer months one sees oddly-shaped wicker baskets. With large openings at the top they narrow towards the bottom. Fish bait is placed inside and when a lobster crawls through the neck it is trapped, unable to back out. The baskets are emptied early in the morning, and at night lowered to the sea bottom. Sometimes they remain there for days before a lobster, a crab or another shellfish happens along.

Golubica had just put on a pot of dried beans for soup. She said that the summers were generally hot and dry, but that during the winter enough rain water collected in the cisterns to last the summer, so that they rarely ran out of water. Pikolo made no comment; he did not even look up from his nets.

They didn't much like having strangers on the island, especially underwater fishermen whom they called poachers, although they all had duly issued permits. Every so often a yacht would sail into Pikolo's cove (a good portion of the island was his) and anchor there for several days. The old fisherman gladly shared his catch with the crew and sold them wine, of which there was always a surplus. We found a German yacht anchored off Lavsa; its owner was from Munich. Words failed to describe adequately how they felt about their stay on the island. They ate fish, washed down with copious amounts of wine, at a stone table, made friends with their host, and, at his guests' insistence, Andrija Juraga, who had never in his life ventured further than the island of Kornat, promised to visit them in Munich. The yacht with its female figurehead was suitably named Calypso. The absence of clothing on the deeply tanned bodies of his visitors didn't bother the old fisherman, no longer startled by the whims of holiday nomads. If there is anywhere in Europe, they said, an oasis where one can escape the fetters of civilisation and live freely in nature's environment, it's the wild, untouched Kornati isles. "Half a day's drive from Munich," said the Calypso's owner, "and we're in a completely different world," raising his glass of red wine.

Everyone has dreamt of spending a month wandering in the path of Ulysses, discovering sleepy coves, guided by a fair nymph who knew the Adriatic well. The young sun-worshippers made one think of the nymphs of mythology wearing golden girdles, their blond tresses falling to their thighs.

Throughout the centuries men have responded to the call of the sea and set off along watery routes to the unknown. From the Lošinj sea-captains and Ulcinj corsairs, the Uskoks of Senj and other well-known navigators from this coast, down to the commanders of today's ocean liners — all have felt the irresistible attraction of the sea, the desire to grapple with the fury of its waves and tempests. Since ancient times men have dwelled near the sea and on it, offering a timeworn Latin phrase to explain why men go to sea: *Navigare necesse est.*

MURTER

85. *A scene typical of many island communities. The old woman working in the courtyard, probably the wife of a fisherman or peasant, is surrounded by objects she has used all her life. Dressed in black as custom dictates, these women are symbolic of an earlier way of life and a patriarchal society, of the honesty and unflagging optimism of the Adriatic people.*

86. *A fishing harbour on the island of Murter. When the fishing fleet returns after several months in the Kornati waters, the idle winter months begin. Now and then the raucous braying of a donkey breaks the silence, a reminder that it is time to eat.*

87. *The shipwrights of Betina on the island of Murter are known for their solidly built wooden boats. Made entirely by hand, each boat is unique and on the sea reacts differently from all other boats. Ship-builders say that boats have a personality, just like people.*

26TJ

61 TJ

MURTER

88. *The fishing village of Murter, on a Central Dalmatian island which bears the same name, developed at the edge of a fertile plain on the northwestern side of the island. Its inhabitants own many of the Kornati islands where they spend the summers fishing and catching lobster. In prehistoric times the island was inhabited by the Illyrians, the first known settlers on the Adriatic coast.*

KRAPANJ

89. Above. *The village of Krapanj, on a small Central Dalmatian island of the same name, has an ancient tradition of coral-diving and sponge-fishing.*

Below. *On the island of Krapanj all the farmwork is done by the women, for the men are away at sea diving for coral for long periods. The village has a sponge factory, and the women polish the coral to make jewelry. In the summertime figs are dried at the water's edge.*

90-91. *The isle of Krapanj practically hugs the mainland coast, closing the southeastern end of the Šibenik channel. Only 300 metres of water separate the island and its village from the mainland.*

92. *Far out at sea a ghostly apparition heaves into sight. Sails billowing, the tall ship cleaves through the waves under full rigging. Tales of free-booting corsairs spring to mind. Actually this is a naval training ship where future captains master time-honoured maritime skills.*

Nowadays the call of the sea attracts people who cannot claim it as an occupation. A sojourn on the sea, on its gentle shores — *necesse est* — because it gives pleasure, it relaxes and brings one close to nature, to what men were before civilisation. The Kornati isles! What an opportunity, at least for a while, to live the life of Robinson Crusoe, or at any rate to enjoy the illusion that life on these desolate islands is a little like that of Robinson Crusoe. Andrija Juraga was cleaning a big squid which he had just caught with his harpoon. He was going to make a "black risotto," sprinkled with cheese, and the tendrils would be cooked and eaten as a cold salad. The Robinsons and nymphs from the Calypso would have a memorable lunch and another tale to tell about their enchanting desert island. We watched them raise anchor and set off for a morning's sail amongst the islands, until Andrija had got lunch ready.

We sailed on, the sun high, and the waves, whipped up by the wind, lifting the boat as if to dash it against the steep rocky cliffs. Seagulls circled above, their nests high on the cliff tops, inaccessible and well concealed from human hands. Sailing among the islands one gains the impression that this desolation supports no life and never has. No great armies clashed here, nor did bishops and princes build palaces. Nevertheless, the Kornati isles have a history, the history of ordinary people: fishermen, shepherds and labourers since time immemorial have struggled against uneven odds simply to secure a bare living. It is an unwritten history beginning in the neolithic era. Late Stone Age men, whose artefacts have been found on the island of Piškera, knew how to construct a boat or raft and sailed from island to island. Illyrian tribes also settled on the islands. The Illyrian wall foundations, fortifications and burial sites are two thousand years old. The Romans, however, did the most building. Near Mala Proversa we sailed through a shallow channel; beneath the water lay the remains of a Roman villa. Spreading across an isthmus on the island of Piškera were the walls of a Roman toll-station. Soldiers guarded this fort which controlled the approach to the islands; fishing rights were paid in kind. During the Middle Ages the islands offered refuge to people from the mainland fleeing attack and pillage.

We disembarked at Piškera, an island as rocky and parched as the others. A stone wall scrambling up a hillside across the middle of the island indicated that the property had two different owners. Milan pointed out some houses roofed with stone tiles. In the summer of 1617 the Uskoks of Senj set fire to the settlement and took 150 prisoners. The Uskoks, who guarded their home city Senj from enemy attack but also harried Venetian shipping, were on good terms with the Kornati fishermen and never interfered with their fishing. That year, however, they had learned of a Venetian galley which had come to the island with merchants who wanted to buy fish. About 160 Uskoks, armed to the teeth, landed and captured the trading party and the sailors. They executed eighteen Venetian merchants and carried off the others. Among them were a few Kornati fishermen.

A short distance from the shore, alone and exposed to the scorching sun, stands a church built in 1560. During World War II it served as a field hospital sheltering wounded Partisans who had been transported from the interior, from as far as Bosnia, during the fierce fighting known as the Fifth Offensive. Here their wounds were freshly bandaged, and after a brief rest they were taken by boat to the Partisan-held island of Vis and from there by Allied planes to Bari and the hospitals of northern Africa. Near the church, right at the water's edge, is a monument and crypt containing the remains of those who died of their wounds en route or while guarding the hospital. Thus these tiny deserted islands have a place in the pages of recent history.

Our boat continued south round Piškera and onto the open sea. With the waves rolling high we sailed in as close to the steep shores as we dared, close enough, however, to hear the waves breaking against the soaring cliffs of Lavsa and Levrnaka, limestone rock chiseled into bizarre formations under the abrasive action of the waves. Beneath Klobučar, its sheer cliffs rising 99 metres straight out of the sea, the waves thundered dangerously, powerful enough to smash a small boat to smithereens. Before the days of the gyrocompass and radar, many ships driven by high seas or wandering off their course in the fog met their end here. The blue grave was deep: the rocks drop more than 100 metres to the sea bed.

A surfeit of beauty can also be wearying. We are afraid of receiving too many impressions and looked up to see if the sun was beginning its descent. We sailed through the Proversa channel past the oddly sited *Kod Mare* restaurant where one can be sure of getting good fish. Its only access is by sea; behind the building stretches a bare, rocky expanse. A few fishing boats were moored at the jetty, and we were invited to share some good homemade wine, but we were in a hurry. We had to cross Murter Sea, a good two hours' sail. We left the channel and after passing the tip of Dugi, the nearest island to the Kornati, we headed southeast, towards Murter. Many fishing boats crossed our path. People have fished in these waters since Roman times, and throughout the Middle Ages they were considered the best fishing grounds in the Adriatic. Zadar's patrician families fought for their control. As our boat chugged on towards home port, Milan described different fishing techniques: with a *parangal*, a long rope fixed with a dozen hooks and trailed on the surface of the water; with a *panula*, a baited hook drawn behind a rowing boat: by *sviću*, fishing by night with acetylene lamps and harpoons. The last technique is also called *na osti*. In spring they fish *na koćare*: powerful motor boats drag large nets over the sea bed. There are also wire baskets called *vrše* which are lowered to the bottom of the sea early in the morning, and drawn up, later in the day, generally filled with fish.

What kinds of fish inhabit these waters? Milan listed as many as he could think of: sardines, anchovies, mackerel, pickerel, grey mullet, hake, red mullet, dentex, bream, all kinds of shellfish from the tiniest crab to the big lobster.

The life of ordinary fishermen was very hard. They were called *tovariše*, a name which has little connection with the Russian word for comrade since it comes instead from the word *tovar* which means burden; in other words, "one who bears a burden." The fishermen pulled nets, pulled oars for long hours from one island to the next. Until quite recently their manner of life never changed. So long as the fishermen had only oars and sails, they necessarily spent more time on the Kornati isles than they do now that their boats are powered by engines.

The setting sun poised for a moment on the horizon. We

glanced back at the islands reddening in the sunset, scattered over a golden sea. As in a dream, the day's experience, the visions of harsh beauty flicker through one's memory, fleeting, unreal, belonging to another world.

Sailors have always brought home souvenirs from the countries they visit. This is how Korčula acquired this 15th-century Madonna and Child *from the island of Crete. The Cretan school of icon painters were influenced by Byzantine artistic traditions. (From the Cretan Icon Collection in the All Saints guildhall in Korčula.)*

MURTER
An ancient craft of shipbuilding

Hammers were chipping away at the curved wooden ribs of a boat in the making. We were in the Betina shipyards on Murter island. Small docks, propped up by beams, supported the skeletons of future fishing boats and pleasure craft. Fishermen have little faith in plastic and mass-production. They want to hear the reassuring sound of water breaking over a wooden prow. Owners of much more expensive crafts come to Betina or some other old shipyard to place orders for wooden boats. One at a time the hand-hewn strips of wood are bent over fire and doused in boiling water. This ancient craft of shipbuilding is practiced much as it would be along the Amazon or the Blue Nile. Time has brought the techniques to perfection. These traditionally shaped fishing boats, as well as the gracefully modelled sailboats, will bear the mark of one of the Adriatic's best known shipyards.

Entering the workshop of an old shipwright we saw, hanging on the walls, designs for ships and boats of all shapes and sizes, big wooden compasses, row upon row of chisels, hammers and saws. This is where the boats are conceived, the all-important keel made entirely by hand. A boat's seaworthiness – whether it turns out to be a *vergul*, as old Jure said, liable to capsize, or rides the waves easily – depends on how the ribs are bent and joined. Betina's boats are known to be solidly built, durable and exceptionally seaworthy. Formulas guaranteeing this kind of quality do not exist, or at least they are not made public. This is not assembly-line production; each boat is built by hand, each with unique characteristics even though they may all have similar dimensions and design. Talking about the big fishing boat on one of the docks, the builder spoke as if he were about to breathe life into it. He knew how it would behave on the sea, how fast it would sail, how much cargo it could carry.

The old craftsman lived with his boats, as long as they were on land. Once they were launched, if they remained at Murter, he kept a paternal eye on them. We asked him what he earned. Money isn't important, he answered, enough to live on respectably as an artisan. "It's better to lack money than friends," he said with the wisdom of his years. From the old builder we heard a number of sayings and proverbs that still flavour village conversation. One of them, sounding quite anachronistic now, refers to a time when many people left these arid, rocky parts for the Americas: "It's better to have an enemy for a neighbour than a brother abroad."

The Kornati isles were hidden beneath a dark coverlet. Flickering in the distance was the Sestrice lighthouse, its beams flashing over the sea, a star plunged into the water...

In addition to Renaissance and Baroque paintings by Italian and native artists, a collection of gold-embroidered church vestments, the Abbey treasury in Korčula also possesses a lapidarium with stone fragments dating from the 15th to the 18th centuries. This relief depicting Christ's Passion (15th century) is carved in English alabaster.

Sailing South

MLJET
INDIAN
MONGOOSES
AND SEA-LIONS

The mingled scent of Aleppo pines, myrtle and wormwood; ribboning white-edged coves and inlets, silent and promising; a bower of pine-needle branches bending to the water, rippling at the touch; and across a wide channel the towering mountain barrier of the mainland – we were on Mljet, an island where every path holds promises of wonder and delight. In every cove the air is filled with the fragrance of wooded shores. Over two thirds of the island is covered with pine forests, oak and scrub. On the remaining third there are three fascinating lakes joined by a channel to the sea, a lake isle with a Benedictine monastery, a dozen villages on the hillsides, and a wide flat valley known as Babino Polje (Old Woman's Valley). The valley contains Babino Selo (Old Woman's Village). I asked some peasants at Maranovići, where one of the paths had taken me, about the origin of these names, what old woman had lived there. They replied that once upon a time there had been an old woman, a beauty in her youth, who had a palace in which she never resided, preferring to dwell in a cave in the midst of the loveliest of gardens. The Romans named the island Melita in her honour. The Greeks, however, had called it Ogygia.

The cave still exists. Known as Galičnik, it is about fifteen minutes from Maranovići. Looking out towards the channel we located a cove, one of several on the northern wooded side of the island. This was Porto Kamara, a sheltered harbour so warm and cozy that the word for "chamber" seemed suited. At anchor in the cove was a yacht flying Panamanian colours, its carefree passengers splashing in the limpid water. It was upon these shores, as legend has it, that Ulysses was cast by a tempestuous sea, after a shipwreck in which his whole crew had been lost. The tradition connecting Homer's hero with Mljet is a strong one. It is the only island in the Adriatic where one finds a cavern matching in description the grotto inhabited by Calypso, the "old woman" of whom the islanders still spin tales.

Ogygia, as the Greeks knew it, is the green-forested island where Ulysses spent seven years in the clutches of the nymph Calypso. Judging from the description that Zeus's envoy Hermes gives of his journey from Mount Olympus to Ogygia, the island must be in the outer string of the south Dalmatian archipelago, one of the last larger islands near the shore, from which one could quickly reach Scherie, today Corfu. The island was inhabited by the nymph Calypso, and the names of the valley and village suggest that the immortal Calypso grew old on the island. The "old woman," according to the islanders, ruled the island and had a palace at Babino Polje.

We climbed a narrow path until in the thick underbrush we suddenly spotted the dark opening of the Galičnik cave which corresponds so closely to Homer's description. Standing at the entrance I seemed to hear the voice of the legendary Ulysses... The spacious cavern opening at the base of steep rocks overlooked a gentle valley where grapevines were running riot. On either side of the vineyards stretched poplars and cypresses, and the path to the cave had borders of rosemary and myrtle. I heard the trill of blackbirds and larks; out of the cave flew swallows and doves. The heady beauty of the scene drew me closer to the cavern, dark and moist...

Even today the setting is romantic. Overgrown with low scrub, inhabited by owls and ravens, the cave excites our imagination, and in a flight of fancy we picture Ulysses' first meeting with the nymph Calypso.

... A large stone rolled out of the cave. Someone was approaching. With footsteps echoing in the obscurity came the sound of a voice, strong and low, yet feminine. The figure of a woman appeared. She was beautiful, her dark bronzed face framed in fair locks that tumbled over her shoulders revealing strong, well-formed arms. A thin, transparent veil, gathered about a slender waist with a golden belt, outlined a body of feline grace...

Calypso, child of Atlas, mistress of Ogygia.

Ulysses king of Ithaca, lost hero of the Trojan War.

LASTOVO

101. *Solitary in its beauty, this island belongs to a group of isles scattered round Lastovo. Though covered with the thick vegetation which is characteristic of this distant archipelago, the island is not inhabited.*

VIS

102-103. *Lying far out at sea is the island of Vis, its past punctuated by a series of great naval battles. Its possession was disputed by many powers from the Byzantine Empire to the Venetian Republic. In the mid-19th century the Austrian fleet defeated an Italian force in a naval battle off its shores. For a time during World War II the headquarters of the Yugoslav People's Liberation Army was based on the island.*

104. *The harbour of Komiža, centre of the island's fish-canning industry, lies in a sheltered sandy bay on the western end of Vis. Benedictine monks moved here from the neighbouring island of Biševo in the 13th century, but their monastery dates from the 17th.*

105. Above. *The Blue Grotto (Modra Špilja) is located in Balun cove on the island of Biševo. When the sun is high and the sea calm, sunlight penetrates cracks in the cavern walls and is reflected from the rocky seabed in an azure blue light which has given the grotto its name.*

Below. *The fertile valleys of Vis are dotted with vineyards. Thriving on the sandy red soil, the vines yield a heady red Opolo.*

VJESNIK

CAV
ANTONIO
SANGALETTI
PER SE
E FAMIGLIA

They spent seven years together.

I descended a steep path through pine woods to the beach of Sapjunara. On the soft sand flanked by ancient forests Ulysses had built a raft which, with the goodwill and counsel of his mistress, had carried him to the shores of Scherie (Corfu) and a new adventure with the lovely Nausicaa, closer to his native Ithaca. Mljet's tall pine forest aided Ulysses on his return after years of wandering among the unknown islands of the Adriatic. On Mljet the homeless wanderer had encountered a woman graced with divine beauty and human frailty. Tender and wild, she bound him to her with feminine wiles – yet remained an enigma. The daughter of Atlas was a true nymph, ardent and demanding. Yet one day when the gods on Olympus commanded it, Calypso took her reluctant lover by the hand and led him down to the sea where the trees grew tallest. She gave him a giant axe and pointed out a century-old fir, its branches reaching to the sand. From the tall trees Ulysses put together a raft with his own hands and set sail, escaping a prison of passion and oblivion.

Centuries after Ulysses and Calypso, others were consigned to oblivion on the island. The Roman poet Agesilaus was punished by his emperor Septimius Severus with exile on Veliko Jezero, one of the three lakes which have an exit to the sea. The ebb and flow of the tides is so strong that water is constantly pouring either into the lakes or out of them. On the banks of the channel Benedictine monks built mills; the remains of these mills are still visible. The monastery itself, built in the eleventh century on an island in the middle of the lake, has survived intact; it is now converted into a hotel. Alongside the monastery is a Romanesque church dedicated to Our Lady of the Lake. The monastery walls shelter fragrant rose bushes, orange and lemon trees, a gurgling stream, its water cool and refreshing.

As the exiled poet Agesilaus wrote to the emperor: "In a land where birds sing on the green branches of a pine, where the sea frolics with the shores of a cove trimmed with dark fir trees, through which the western sun casts shafts of gold on the glittering water, there can be no evil, no sorrow..."

The poet lived out his days far from evil and sorrow. After him came Venetian aristocrats who built summer

LASTOVO

106. *Ribboning sandy-edged coves, the soft branches of fir trees almost touching the water, the scent of Aleppo pine hanging upon the air – this would describe any of the verdant isles gathered round Lastovo, about 13 km. from the island of Korčula. The tranquil coves offer anchorage and a safe haven for boats on stormy days at sea.*

107. Above. *Inside the home of Kuzma Katić, an aged fisherman from the village of Zlakopatica on Lastovo. Round an oil painting of his mother in her youth, the old man has arranged a collection of magazine clippings: movie stars, front-page figures, whatever caught his interest.*

Below. *Old Kuzma sitting, as usual, at his window, the first to notice a stranger in the village and start up a conversation. The sleepy hamlet offers a scene of unexpected beauty: white houses clustering round a cove where fishing boats bob up and down on the translucent sea.*

108. *The small town of Lastovo on an island bearing the same name. From the cemetery and its weathered gravestones one has a view of ancient houses which never echo sounds of youthful laughter. The young people are deserting the islands for jobs on the mainland.*

residences hidden in luxuriant vegetation and sweet-scented flowers. The crumbling walls of these palaces, lost in ivy and overgrowth, conjure up a time when the island held promises of sunshine and oblivion.

Along the rocky southern coast of the island one may have the rare experience of observing sea-lions basking with their cubs in the early morning sun. Their round wet heads with mournful eyes gleam in the sunlight, but it is a sight reserved for early risers. The sea-lions swim to Mljet from the distant islands of Palagruža, Jabuka or Biševo, and because of their eyes and gentle expression they are called by the islanders "men of the sea." These are the only specimens extant in the Mediterranean. Scientists estimate their number at thirty pairs and steps have been taken to preserve them. They have been killed off by fishermen because of the damage they are believed to do to their nets. There is even a belief that at night the sea-lions emerge from the sea to trample the vineyards. This rare animal is also blamed for harm done by other animals, for example dolphins, who actually do tear nets and steal fish.

A much better known inhabitant of Mljet also belongs to the animal kingdom. Its presence in the island woods began one day early in the century when a local sailor thought of a way to help his island. In those days the rocks and woods were infested with poisonous snakes. They made life very difficult for the peasants and shepherds, especially when they were felling timber or cultivating their vines. A bite could be mortal. One of the island's many sea-captains learned that in India an animal called the mongoose had a great appetite for snakes. Swift and cunning, the mongoose was unusually keen on hunting snakes. So one winter's day a ship sailed into the harbour of Polače with a pair of mongooses on board. So diligent were they, and not only at catching snakes, that in a few years there were so many mongooses that all the snakes had been killed. But they kept on multiplying. At first the inhabitants of Prožur, Maranovići, Blato and Babino Polje rather liked the squirrel-like little animals that so bravely attacked and choked the snakes. But the snakes disappeared, and there were more and more mongooses. They started to feed on chickens, ducks and other poultry, slipping into hencoops and strangling the baby chicks. The peasants were soon obliged to set traps for the newcomer from India. During the nights the mongooses can be heard yapping in the forests, not about to abandon their own particular struggle for survival.

Night steals quietly over the island forests. Under a star-studded sky day surrenders to night with the fading whir of crickets; their rhythmic monotonous song is heard throughout the summer nights.

Trained in Venetian Gothic, Trifun Kokoljić (1661-1713) did most of the paintings in the Church of Gospa od Škrpjela, situated on an islet in the Bay of Kotor. Other paintings bearing his signature are in the Dominican monastery of Brač and in the Benedictine monastery of Mljet.

Standing beneath the portal of the Benedictine monastery one wonders at the choice of this site as a place for meditation. It was given to the monks in 1151 by Prince Desa of Zahumlje; the monks came all the way from Pulsano abbey on Mt. Gargano in Apulia to build a monastery and church in the middle of this lake. The Benedictines no longer inhabit this island; they form part of the past as recorded on weathered gravestones. Prizing a hold in the cracks of the stone portal was a fragrant caper bush, its ripening berries dropping tear-like to the worn stone flagging. Inside this monastery there are paintings by Titian, Bassano, Veronese, Trifun Kokoljić and Celestin Medović.

Stepping further into the past we reach Polače where vineyards and olive groves surround the ruins of a fortified Roman villa. The rose-tinted patina of the Dalmatian stone is set off to advantage against a silvery bouquet of olive branches. According to tradition the poet Oppian wrote his *Cynegetica* and *Halieutica*, poems about shooting and fishing, in the solitude of the island landscape.

We thought of Sapjunara, the waves whispering on the soft sand. A true island scene — the long line of the horizon where sea and sky meet, the wind humming through pines, needles dropping gently on the hard stony ground, waiting to cushion the weary. We thought again of the bronze bell tolling from the monastery church, its vibrating peal scattering the swallows at sundown. We remembered the flowering caper bush surviving in a crevice between two stone blocks. From the cupola above Bacchus laughed, bees cautiously descending his grey stone beard to the flowering bush. Gonoturska cove lay beneath night's mantle in the most profound silence — an old monastery, a church and the clouds.

Homer's description of Mljet is of an island of unimaginable beauty. He was only repeating the words of travelling minstrels. From their lips Ogygia sounded like a lotus flower. Sailors had recounted tales of faraway Ogygia, where the nymph had offered Ulysses eternal youth if only he would stay. The sailors spoke of the high rocky shores of the Hylide peninsula, where Hercules was born. Hylide — these were the rocks of Pelješac, just across the channel from Mljet. Tales summon up visions, but how much better it is to go there oneself, an experience to be cherished for a lifetime.

VIS
THE MALTA OF THE ADRIATIC

We were already at a good distance from the coast. In the steamer's frothy wake lay ships of stone anchored firmly to the Adriatic seabed. We were ploughing towards a new acquaintance — distant, lonely Vis. Seagulls circled tirelessly behind the flag that fluttered from the stern, swooping every now and then to snatch up a piece of bread from the water. The further we got from land, the greater their numbers. For a few moments they would float in the trough of a wave and drop behind the fast steamer. Taking wing they would catch up, sharp eyes watching for the tiny bits tossed from the deck. Not one attempted to land on the ship and break its graceful independent flight.

Early in the last century (1815) the Congress of Vienna awarded Vis to England. Vis town was named Hoste after Captain William Hoste who led the English forces in a naval battle just off its shores. Three hills near the town also bear English names — Bentinck, Robertson and Wellington. As fortifications the English installed cannon embrasures in towers built into the town walls; the walls themselves rest on ancient foundations. The "Malta of the Adriatic," thus named because of its strategic location, has been close to the scene of battle on many occasions, the last quite recently. During the last two years of World War II it was a naval base serving the People's Liberation Army of Yugoslavia, "the Partisans' island" as it was called. In June 1944 it became the wartime capital of a new Yugoslavia, headquarters for the Partisan army and the National Committee for the Liberation of Yugoslavia which was acting then as a provisional government.

The Greek colonies on the Adriatic islands produced distant echoes of Grecian art. Terracotta vases of various shapes reveal skillful, elegant workmanship. A vase from the Greek colony on the island of Issa (Vis).

Marshal Tito spent time there before the capital, Belgrade, was liberated. RAF planes landed on an improvised military field, and British patrol boats docked in the harbour of Komiža on the western end of the island. Thus once again the Malta of the Adriatic – which had entered naval history with a battle just off its shores on July 20, 1866 when the Italian fleet was routed by the Austrians, with Admiral Tegethoff in command – had its share in the struggle for supremacy in the Adriatic, and late in 1944, together with the Allies, celebrated victory. On two separate occasions there were English troops on the island – once to occupy it during the Napoleonic period, and a second time as allies.

Right on the waterfront of Vis harbour stands a little café which offers a choice of three local wines: the dark red Opolo, the sweet golden Vugava, and Plavac, a heady red wine. On Vis wine-fanciers can add to their collection of original vintages from the Adriatic vineyards. The same goes for fish. After a moonless night one's plate may contain grilled mackerel or sardines brushed with pungent olive oil. If moonlight has kept the fishing boats at their moorings, one of the sunken fishing baskets may produce a good hake or a bream. Fish stands foremost in island gastronomy. Nevertheless, the stone hearths of the island villages produce other delectables. Bread, for example, is baked *ispod sača*, under a tin cover piled high with hot ashes. The women themselves grind the breadflour. Yellow cornmeal is made into a thick *polenta*, eaten either with fish or with milk. When Bernard Shaw in his travels about the islands tasted a local *brodet*, his comment was – "the poetry of the Adriatic coast, cooked in a pot." The secret of this cuisine lies in the seasoning: a variety of aromatic Mediterranean herbs. Then add to your repast a sun-steeped Dalmatian wine, or if you like weakened to a *bevanda.* Wary of the combined effects of strong wine and a strong sun at noontime, Dalmatians drink their wine diluted with water. As an apéritif, or any other time, there is the Dalmatian *prošek*, a brandy distilled from wine. And if your squid risotto is served by pretty hands, why so much the better!

The islands have inspired many a poet. Dionysius while sailing the Adriatic found the islands an endless source of themes for drama and tragedy. Torquato Tasso was so impressed by their beauty he thought them a fit habitat for Aphrodite. Herman Bahr, as a more modern representative, had never seen a land where life held so much poetry and beauty as the island-strewn Adriatic.

But the attractions of Vis do not end with food and wine, or the romance of sandy shores. On the stone terraces of sea-girt Vis, ancient Issa, we find the remains of first-century villas and mosaics. These were thermal baths; and on the Prirovo peninsula, the remains of a Roman theatre which in its turn was built on the ruins of a Greek theatre. Two theatres, two eras — Greek and Roman — on a distant Adriatic isle. There, too, are medieval churches watching over the town walls — a picture that is repeated from island to island, a picture that reflects a good part of the history of European civilisation. Looking down from a side aisle in Our Lady of Spilica is a fine *Madonna with Saints*, a fifteenth-century Renaissance painting by Girolamo da Santacroce. The decoration of sixteenth-century St. Cyprian's, its wooden pulpit and painted coffered ceiling, bear a stong Baroque imprint.

This detail is part of a polyptych painted by the Italian artist Girolamo da Santacroce, a pupil of Bellini, for the Church of Our Lady of Spilice (built in 1500) in Vis.

This beleaguered island with its romantic landscape, just waiting for some poet or painter, lives under the sign of peace, in the shade of palm trees. Groves of palms cluster along steep rocky shores; and in the gardens are orange and lemon trees, eucalyptus, mimosas and flowering cactus. This is the south, an island swept by hot winds. The narcissus blooms in January, almond trees in February, cherry trees in March.

BIŠEVO, SVETAC AND BRUSNIK

THREE SMALL ISLETS AROUND VIS

Vis also has its offspring, smaller replicas of itself. The closest is Biševo. Here, too, we must turn the clock back. The Benedictines, whose monasteries occupy some of the loveliest sites on these islands, arrived in the eleventh century and built the Church of St. Silvester. In the twelfth century they were attacked by pirates from Omiš, and in the same century Turks set fire to the pine woods, the monastery and its church. Many other islands shared a similar fate. Some two hundred fishermen and vintners live on the tiny island. Some sell wine, others fish. Yet they all live together in harmony. On Sundays they row together to Vis, the farthest they ever go, to visit friends. Thousands, however, come to visit their little island. In Balun cove is the Modra Špilja (Blue Grotto), sister to the one at Capri. One enters by boat, the sunlight reflected through the water tinting the cavern walls an azure blue. The white rocks beneath the green water glimmer in the incandescent light, making it a memorable experience.

Drifting over an unknown sea one sunny day Ulysses came upon Biševo, the floating island, as Homer poetically describes it. This appearance of floating is quite common in the Adriatic, especially during very sunny periods. Due to the refraction of the sun's rays, like a mirage in the desert, the island shores seem eaten away by the sea and, from a distance, the island appears to be floating above the surface of the water.

Biševo or Palagruža may have been the home of Aeolus, Greek god of the winds. To help Ulysses homewards, Aeolus imprisoned all the winds in a leather bag which was stowed aboard Ulysses' ship. His men however, thought that the bag contained gold and when Ulysses dropped off to sleep they undid the bag and out rushed the winds, blowing up such a tempest that his whole fleet was driven back to the unknown waters of the Adriatic.

Thirteen sea-miles from Vis is the island of Svetac, or Sveti Andrija: a pile of rocks topped with clumps of pines. According to tradition the Romans and Illyrians used it as a place of exile. On the southern part of the island are ruined fortifications that belonged to the Illyrian queen Teuta.

Fishermen from Vis and Svetac catch lobsters in the waters of lonely Brusnik. Near it is Jabuka with its coral reefs, ideal hiding-places for lobsters. Whenever ships approached this trio of islands the compass needle would no longer point north. The reason wasn't too difficult to discover: all three islands have a high percentage of iron ore in their rocks.

LASTOVO
THE GREENEST ISLAND IN THE ADRIATIC

Far out to sea, like Vis and its satellite islands, lies the Lastovo archipelago. The Roman historian Pliny the Elder called this group of verdant islands Celadus, and the Greeks before him Ladesta. The Byzantine emperor Constantine Porphyrogenitus knew it as Lastobonum. To Croatian settlers since medieval times it has been Lastovo. Ruled by a succession of masters, Lastovo enjoyed its longest period of freedom as part of the Republic of Dubrovnik, from 1252 until 1818. Yet even in that Slavonic union the people of Lastovo aspired to autonomy and their own statute of privileges.

During a period of rebellion the Dubrovnik governor convoked some influential citizens of Lastovo to learn why there was so much trouble and discontent on their island. The Lastovo delegation cleverly proposed an exchange of favors: "You give us freedom, and we'll give you trust." The governor was wise enough to accept this laconic offer, and the people of Lastovo settled down peacefully; in return the governor gave them a statute guaranteeing autonomy.

I heard this detail from Lastovo's eventful past while standing in the main square from which one has a fine view of the town and its roofs which rise in tiered layers up a rocky hillside. Sitting in the Loggia, where the town council used to convene and where criminals were tried, was a group of older men: retired sea-captains, fishermen, peasants, talking as usual about the good old days. One of them mentioned the number 46 as having special significance for Lastovo. So convinced are they of the number's special meaning that they even bet on it in the national lottery. The old men explained. The Lastovo archipelago consists of exactly 46 islands. On the largest there are 46 fields planted with vineyards, and according to the islanders they have counted 46 hills, each bearing a name. The repetition of the number is surely only a coincidence, but if one looks doubtful one is invited to go count the islands, fields and hills for oneself. While the old folk talked, chuckling occasionally at some witticism, my glance was drawn to the old roofs and a forest of chimneys. Native fantasy has run riot with these practical, necessary appendages without which no fire would burn in the hearths. The taller ones resemble slender minarets; the smaller ones, the turbans of Moslem dervishes,

117. *At the end of every journey one seeks the intimacy of nature, the restful shade of a venerable pine, the chirp of birdsong. As the sun disappears behind the hills of a nearby island and casts shadows over the landscape, an island holiday ends with a scene that is long remembered.*

118-119. *Three "stone ships", three humps of sea-girt land. In a tumultuous operation lasting thousands of years ancient tectonic forces broke up the Adriatic coastland to produce the winding shores and strings of islands that we admire today.*

all of them round and pointed, built of stone and mortar. When the fierce winter winds spin smoke over the roofs, the chimney-pots look like busy chefs hovering over the cauldrons of a field kitchen.

As with most of the islands, Lastovo's name has a story. As the fearless, belligerent Slavs pushed westwards, they dislodged the Roman population from the islands. By the time of Constantine Porphyrogenitus Croats from the Neretva River, well-known corsairs throughout the Middle Ages, had settled on Lastovo. This occurred in the seventh century. In time the Latin names for the islands were modified to suit Slavonic tongues. Whether this was the case with Lastovo, or whether its name is of purely Slavonic origin is a matter of conjecture. At any rate, the old men of Lastovo insisted that the island got its name because from its highest point it resembles a swallow's nest (*lasta* meaning swallow), and the surrounding isles, swallows. Thus the Croato-Serbian derivation of Lastovo would be "swallow's nest."

The tales spin on in the lengthening shadows of the dormant chimney-pots, not in use on hot summer days. The loveliest ornament of the island, however, is its vegetation. Lastovo is one of the greenest islands in the Adriatic, rivalling even Mljet which is famous for its flora. Forests of Aleppo and Lebanon pine, black oak, scrub, heather and juniper deck the island in verdant luxuriance. The forests reach down to the shores, and where they leave off, the blue sea begins. The old folk claim that the islands to the south are more verdant than the northern ones because the Republic of Dubrovnik had stringent laws protecting the forests. For five centuries it was customary for any young couple who wished to marry to plant an olive tree, a carob bush and a fir tree. There were lots of weddings and wedding-feasts, said the old men. This is one custom one would like to see restored.

Accompanied by the parish priest we set off to the Church of St. Cosmas and St. Damian. Removing their caps respectfully, the old men seated themselves in the pews. They stopped talking in deference to the priest. The church was built onto a fourteenth-century chapel which forms the walls and vaulting of the main altar; the Gothic nave was added later. At the entrance to the church on a stone pillar

PALAGRUŽA

120-121. *Palagruža — a tiny cluster of isles and reefs far from the mainland, almost in the middle of the Adriatic, about halfway between Yugoslavia and Italy. When fierce winds blow, the reefs of Palagruža become dangerous, and its lighthouse is of great help to sailors.*

122. *When thundering waves batter against the rocks beneath the lighthouse, the tiny island seems to rock like a boat. Palagruža is uninhabited except for the lighthouse keeper's family who live here alone.*

123. *The lighthouse-keepers of Palagruža, lonely and faithful to their task, have maintained this tradition for over a hundred years. Even today with radar and modern electronic devices navigation would not be safe without lighthouses to warn sailors of unseen dangers.*

124. *Boats are beached on a sandy strand beneath the lighthouse since Palagruža has no landing-stage. In earlier times, when fishermen lived on the island, their wives used to wash their laundry in the sea; rainwater was too precious and reserved for drinking.*

stands a fifteenth-century bronze urn for holy water, presented to the church by Dobrić Dobričević, a native of Lastovo who became a well-known printer in Dubrovnik and Rome, signing himself in a Latinized version of his name — Boninus de Boninus.

The town is dying. The chimneys stand straight enough, but few young people walk the streets. This is typical of many places on the islands. Those who remain wait for letters from Australia and dream of passports, ocean-liners, and unknown lands. They know little of the bitter experience of those who left, who in their youth worked on the American transcontinental railway, or deep in the Canadian mines. The world holds out inviting hands, and youth is ever restless, eager to explore the unknown. But if the sea seems too uncertain, or Australia too far, the mainland is only an hour and a half away by a steamer which leaves from Ubli at dawn every morning for Split and the attractions of that busy port, a chance to learn a trade or continue schooling. Faint ringing from Our Lady of the Fields, a church at the foot of the old town, announces the departure of another close friend or relative. A few drops of holy water sprinkled with a twig of rosemary, a handful of earth cast into the open grave so the departed will rest more easily — these customs, too, are disappearing. New ways are borne to the island in noisy motorcars, speedboats and yachts. The old are being cast aside and forgotten.

Fishing boats were bobbing up and down on the limpid waters of sheltered, sleepy Zaklopatica. It was noon and swelteringly hot; a few cats were napping on the warm quay. At a small window overlooking the waterfront an old man sat wiping his forehead with a red handkerchief. He was delighted to see a strange face below his terrace and immediately wanted to demonstrate his vitality, for he was all of 94, and extremely proud of the fact. He had spent his whole life on Lastovo, never leaving the archipelago. A lonely man who liked company, he had been a fisherman, a lighthouse-keeper and a wireless operator. We entered the house. In the kitchen stood a wooden table, the walls covered with clippings from Zagreb magazines — no bathing beauties, but plenty of lovely smiling faces, and amongst them photographs of cousins who had done military service in the Austro-Hungarian army, pictures of boats and ships, dried olive branches and laurel wreaths. Nearing life's end, still hopeful, old Kuzma Katić wanted to get married. He was only waiting for a bride to turn up. He thought it would be nice if we could be his witnesses. What a wedding that would be! Like in the good old days! His neighbour Katica, who was well in her sixties herself, kept house for him and begged us not to take him seriously. He's wanted to get married all his life, explained Katica, and stayed a bachelor, but he's never stopped hoping.

Flinty with age, Kuzma still had projects and ambitions; another twenty years would be just about right. That a person who has spent all his life in one place should have such a thirst for life seems little less than miraculous. Like everything about him, he was full of vitality, making plans that in most cases would remain unfulfilled dreams. Someday people will pause to admire the ruins of our age just as we stop before ancient portals listening for faint echoes from the past; people have always been the same: anxious to preserve life, and the harder life is, the greater the urge to survive.

Another dawn cast platinum rays over the sea, and we sailed on, the dark misty hump of Lastovo behind us. Three young fellows on board were harmonizing on a Dalmatian melody, to the soft accompaniment of a guitar. They were leaving home and never even glanced back at the island where old Kuzma at the age of 94 was still looking for a bride. The roots of a fig tree, its sweet fruit ripening in autumn, pry deep into the crannies of a stone wall; even a pocket of soil will support life, a life perhaps as zestful and rewarding as on some fertile plain.

We had exchanged the fragrance of pines for the smell of oil and the more satisfying aromas originating in the ship's galley. We were sailing south to other islands; there seemed to be no end of them, so generously had nature endowed this coast. But we were already tired. In a continuum of beauty one cannot feel boredom, but there is fatigue. The sun had risen, a flaming lion roaring in the east — to use the metaphor of a Serbian poet who had been drawn to this area and its salt-laden breezes, its shady grapevines and gnarled olive trees, the soaring cypresses and flowering lemon trees.

In a word — the Mediterranean. The name seems to sum up everything, the name of a region with special features in abundance.

Some time ago in the Veliko Jezero cove of Lastovo, near Pasadur bridge, an ancient amphora, just raised from the sea depths, lay drying in the sun. I had no way of knowing how long it had lain on the sea floor. Almost completely covered with shells, grey layers of stone and seaweed, only small patches of red terracotta were visible. A thousand years ago, maybe more, it had contained wine intended for the crew of some Greek, Roman, pirate or Turkish galley which had been lost in the waves. Sand had swirled round it for many centuries until the galley, and the amphora with it, had become part of the seabed. Now the amphora glistened in the sun, ready to repeat the adventure of wine, sailing and shipwreck.

PALAGRUŽA
Death of a lighthouse keeper

From his lookout on a 90-metre-high clifftop the lighthouse keeper was invariably the first to sight fishing boats homeward bound for the sandy strand of Galijula. Taking a deep breath he would shout down to the women washing their laundry on the beach below: "Palagružians... the men are coming," the men being their husbands, a grand total of three. The three women would immediately leave off work and scramble up the cliff to see for themselves the boats drawing nearer to the island.

There are no longer any fishermen or fishermen's wives. Palagruža's only inhabitants are the keeper, his wife and their two sons. Farthest from the mainland, almost halfway to the Italian coast, this tiny cluster of little islands and reefs is known as an excellent fishing ground. The first of the string is Kamik od Tramuntana, and when a boisterous westerly wind blows up, ships give the jagged reefs a wide berth. Equally dangerous is Kamik od Oštia, though a little closer to the main island of Palagruža. Night and day the beacon from the clifftop flashes its warning light to ships at sea. Keeping the light burning on this isolated spot has been the lifework of men for a long time.

We beached our small boat on the strand, since the

island had no dock, and climbed a steep path and steps to the summit of the cliff. Alongside the tower, painted in stripes, was a stone house, a wall protecting it from the wind and the fury of the sea. If one compares the islands of the Adriatic to anchored ships of stone, the isle of Palagruža is about the size of a dinghy. The whole island, minute as it is, seems to rock in angry squalls. The keeper's home is sparsely furnished: a bed, a stove and a worktable piled with tools. The only other item is the radio-set, his only contact with the outside world. In the late autumn and winter months, when the days are short, he may not venture out for days on end. Outside the sea rages, howling winds drown out the thunder of the surf battering the rocks below. It was on just such a night that the keeper's father died. This lonely duty of lighthouse-keeping is a responsibility passed on from father to son. Without this light navigation in the Adriatic would not be safe, even for ships whose radar screens detect every solid object in their paths.

The present keeper's father, whose picture hangs on a wall near the doorway, used to tell his son about the difficult times of his youth, towards the end of the last century. There was already a lighthouse on Palagruža then, its tall, straight tower marked on all the navigation charts of the Adriatic. In those days it didn't have a code, beams flashing at certain intervals by which ship-captains could identify the lighthouse from a distance. A watch fire was lit on Palagruža's tower only when a ship's bells or fog horn were heard. In his childhood his father had helped his own father build and light the fires. Shuttling two-masters brought wood from the mainland for the island was completely barren. It was worst when the fog descended and closed down the highways of the sea. Then not a soul on board ship took either food or rest. The whole crew stared out into the dense wall of fog. Ringing the ship's bell every minute, the sailor on foreward watch would sound the alarm if there was danger of running aground or colliding with another ship. With clipped, resonant blasts the ship's horn indicated its position in the fog. The lighthouse keeper knew these sounds well, but in a fog a fire was of little avail. The keeper would try to "smell out" an approaching ship, to sense imminent danger. He had a long hunting horn which called out into the fog. A captain's sharp ear would pick up its unfamiliar sound and at the last moment order a change of course. Then he would say with satisfaction to the helmsman: "That was the old fellow on Palagruža." Sometimes they were within shouting distance: "Addio, and good sailing!" Then everyone would heave a sigh of relief: the captain and his crew, the keeper and his family, even the goats seemed to know that tragedy had been averted.

That winter night when the waves were running high and the feverish old man rambled on about watch fires and hunting horns, the keeper radioed to the coast guard station in Split that his father was gravely ill and in urgent need of medical attention. The fierce south wind had reached hurricane strength, and the keeper had lost all hope. Shortly after midnight the radio-set transmitted a message saying that a patrol boat had left Split for Palagruža. Just before dawn another message came through: the patrol boat was leaving Lastovo, the waves having somewhat abated. As the first grey light of dawn touched the clouds hanging low over Stara Vlaka cove, the patrol boat came into sight. After some skilful manoeuvring the doctor and an orderly were put ashore in a sling. The keeper was waiting for them on the strand, but it was too late: the old man had died just as the patrol boat signalled its arrival.

They consigned his body to the tempestuous waves, a final beam sweeping over the water in farewell. The next day the sea round Palagruža was calm, and fishing boats cast their nets beneath the lighthouse rocks.

The Sunny Islands

BRAČ
STONE QUARRIES

In a blinding white glare the sun beat down relentlessly on huge blocks of marble, each weighing several tons. We were in the stone quarries of Brač. The white stone dust was settling on the sweating shoulders of the stone-cutters like a flurry of snowflakes. Even their hair was white. Spread out along paths cut into the sheer quarry walls, they reminded one of extras in some classical play, or the chorus, for example, in Verdi's *Aida* standing next to a pyramid. Instead of a triumphant victory march, a break for lunch was sounded, our only opportunity to see at close hand the men's faces and their strong arms. The island of Brač is famous for its shining white marble which the centuries cannot dull. Near Blaca, above the quarry, is a figure of Hercules carved into the rock. The Romans opened the quarries sending the huge blocks to Rome. Emperor Diocletian commissioned Brač limestone for his splendid palace in Split. According to peasant lore, stone has "always" been quarried here; in other words, ever since men first learned how to dress stone. Stone Age men lived in caves near the village of Donji Humac, and the island's main settlement Supetar: an important neolithic site at the Kopačina cave. The Bronze and Iron Ages also left traces: ruins near Vučja Luka and Gradina, near Bol...

The museum in the Dominican monastery near Bol, a small town on the island of Brač, has a receipt for payment of the Madonna and Child with Saints *commissioned from Tintoretto's workshop, as well as written instructions for the monks who went to fetch the picture. A detail of the Tintoretto altar painting is shown here.*

Brač is the third largest island in the Adriatic. Dotting the rocky hillsides are ruins that bear witness to a stirring past. Dressed and shaped by man, stone survives to relate a varied history.

Brač limestone is exceptionally resistant. Stone-cutters find it hard but not brittle, never tarnishing. Sculptors enjoy working with it, and it is deemed a costly, elegant building material. We were watching as the jagged teeth of a powerful saw cut through a stone block. It was compact, flawless. Healthy stone, said the cutters, many of whom come as a matter of tradition from Pučišća where one of the island's largest quarries is located. There are others at Selca, Postira, Splitska and Donji Humac. The Illyrians and the Greeks used the stone, but its true quality was discovered by the Romans. In the early Middle Ages, when the island was under Byzantine domination, the stone blocks were transported to the East, destined for vast imperial undertakings. From the time of the Hercules carving until the present day stone has figured as an important export item. The Venetians used it prodigiously. The Riva degli Schiavoni and the foundations of the canal palaces have withstood erosion owing to stone blocks from the Brač quarries. The White House in Washington is faced with white Brač stone.

Brač is one of the few larger islands in the Adriatic which does not have a town or administrative and cultural centre bearing the same name. Its ancient name Bratia is derived from the Illyrian word *brentos* (stag). Life in Illyrian times was dependent on hunting, and a killed stag provided meat to eat and skins for clothing. Eventually the deer disappeared; in their stead flocks of sheep ranged over the stony fields. Their summits covered with Aleppo pine, the hills are rocky and barren along their lower reaches. Sheep tug and nibble at clumps of scrub, sharing man's fate in a harsh, unfriendly landscape. The villages of the interior are protected against the summer heat and cold winter winds. Huddling together like flocks of birds, the houses are girded with walls that conceal courtyards, stalls, sheepfolds and small terraces.

It was on one such terrace in the village of Gornji Humac that we were sitting and visiting with Šime Carević. His grandfather had built the house in 1888; he himself had been forced to raise another roof after the last war. Only the walls remained standing after Italian soldiers set fire to the village, known to be harbouring Partisans. After the war few of the inhabitants returned and the younger people were leaving, going to larger settlements on the coast. Over half the village population had emigrated to the Americas. No one wanted to tend sheep any more. The Carević family was dwindling. Šime's father had been killed during the war, his uncle was in America, another brother was a teacher in Split; a third had chosen to stay in the village and tend the flocks.

HVAR

133. *Approaching one of the Adriatic islands from the open sea one often has the impression that the island is surrounded by towering walls. The sheer cliffs seem to gush like geysers from the depths of the sea. The battering waves, tireless sculptors, have carved the limestone rock into bizarre formations.*

134. *On the island of Hvar in late spring and early summer the fields are covered with blue, flowering lavender, its essence used in making French perfume. Another scented oil comes from rosemary, also cultivated on the island.*

135. *Towards the end of June the fields of lavender are harvested. The blue clumps are cut with special scythes and placed in huge cauldrons where the lavender is cooked and distilled. Since the production of lavender oil brings in a good income, lavender cultivation has become popular with the peasants of Hvar.*

136. *Woods and terraced vineyards dot the hillsides of sunny Hvar; only the summits are barren and parched.*

137. Above. *Acetylene lamps for night fishing. Attracted by the powerful light that penetrates deep into the clear water, whole schools of fish are lured to nets which are drawn round them. This method of fishing is an ancient one: in earlier times fishermen held flaming torches to attract the fish.*

Below. *A view of the harbour and roofs of ancient Hvar town. Its mild climate with many sunny days has made Hvar one of the most attractive island resorts in the Adriatic. In the background are the Pakleni isles guarding the entrance to the harbour and sheltering it from the winds.*

138. *The main square flanked by honey-coloured facades. Twin-light windows grace the cathedral's late Renaissance facade. Sun-steeped Hvar radiates the charm of a medieval stage setting executed in stone.*

OBUĆARSKA
RADNJA

Like the last of the Mohicans he lived in a cottage watching over his small flock of 150 sheep. We decided to pay him a visit. The Carević family set a typical example of life in the interior of the island: a peasant living in a deserted house, a valley with untended vineyards, cherry and almond groves. The whole island sustains less than 10,000 sheep; in the past this would have been the size of a single flock.

Jure Carević came out of his smoke-filled cottage, surprised that he should have visitors. He could hardly believe that anyone wanted to see him. The blazing sun had driven the sheep into the shade. Startled by the arrival of strangers, they scattered in the thick brush, their herdsman calling to them in vain. Jure explained that in the summertime the sheep grazed at night, keeping in the shade during the daytime. At sundown he would take them to a watering-place, a dark brown pond of stagnant rainwater. From rainfall to rainfall water collects in hollows and depressions in the stone, the only source of water since there are no natural springs. In the villages precious rainwater is collected in wells called *čatrnje* or *gustirne*.

We returned to Jure's hut. The forty-year-old shepherd offered us some cheese and milk, telling us that it was no longer worthwhile to shear the sheep as wool fetched such a poor price. The work simply didn't pay. He was considering moving down to the coast and taking a job in a hotel. The tourist trade seemed like a good opportunity to leave the village and sheep-raising which apparently had no future. But it was a difficult decision to make. Not every plant thrives in foreign surroundings. Those who left for America were driven by hunger, and the younger ones, dreaming of a motorized urban way of life, had lost interest in the land. For the time being Jure Carević thought he would stay in the village if only to keep it alive.

We left him in his stone hut with his sheep and smoking hearth, and a chorus of tireless crickets. This was, it seemed to me, the proper setting for Loda, shepherd hero of a novel by Vladimir Nazor (1876-1949), native of Brač and well-known Croatian and Yugoslav poet and author. The islands have produced a number of important names in the literature of the South Slavs. Contemporary writers include Ranko Marinković (born 1913 on Vis), short-story writer

HVAR

139. Above. *The 16th-century* Last Supper *in the Franciscan monastery of Hvar was painted by the Venetian artist Matteo Rosselli for the people who saved his life.*

Below. *The facades of old patrician palaces in Hvar. Besides the cathedral and churches blending Gothic, Renaissance and Baroque styles, Hvar has palaces and summer villas built by native master-craftsmen whose skill with stone was transmitted from father to son.*

140-141. *A view of the port of Hvar with the Arsenal to the left. The arsenal was begun in 1579. In 1612 Croatian count Pietro Semitecolo added another story for a municipal theatre which had both plays and opera on its repertoire. The Hvar theatre is the oldest in Yugoslavia.*

142. *On a hill overlooking Hvar a strong fort "Fortica" was built during the Middle Ages. Its cannon guarded the entrance to the harbour.*

143. Above. *On a rise in the middle of Vrboska, a small town on the island of Hvar, stands a fortified church built in 1580. This is one of the few examples of this type of construction in the Adriatic.*

Below. *A 16th-century chapel dedicated to St. John dominates the main square of Jelsa, a small town on the island of Hvar. A Baroque rose-window and bell-tower complete the picturesque setting.*

144. *During the summertime Hvar herdsmen drive their flocks high in the mountains where they build huts to store milk and cheese. Hollows where rainwater has collected serve as watering-holes.*

and playwright; novelist Petar Šegedin (born 1909 on Korčula); Marin Franičević (born 1911 on Hvar), poet and short-story writer. Treating themes from Adriatic life, their poetry and prose probe deep into the mentality of the island people and their response to nature's elemental forces. Their work summons up poetic visions of the barren rocks, and parched, arid soil, vineyards with succulent clusters of grapes, sunsets in snug inlets, the peace of a calm *bonaca* and the distant flickering beacon of a lighthouse, the heady, mingled scent of Mediterranean vegetation, fishermen's songs and fair lasses spinning wool.

The museum in the Dominican monastery on Brač houses a large and fascinating collection of coins which are evidence of the peoples who traded along the Adriatic coast. Emperor Constantine figures on this coin from the museum collection.

It was on the island of Hvar, a thriving centre of medieval culture, that the poet Petar Hektorović composed *Fishing and Fishermen's Complaints* (1555). This Renaissance poem describes the hard life of the island fishermen and in a vigorous poetic dialogue evokes the beauty and wealth of the island. Another Croatian poet, Hanibal Lucić, Hektorović's contemporary, wrote lyric dramas of peasant life; in fact, his folk plays are often performed on public squares at carnival-time. Lucić and Hektorović marked the beginnings of Croatian literature.

HVAR
POETS, FISHERMEN AND LAVENDER

Enjoying 2,700 hours of sunshine yearly, the island of Hvar deserves its title as the Yugoslav Madeira. It is one of the islands most visited by tourists, popular even during its mild winter.

An island that prompts meditation on art and architecture, it leads us on an exciting trip back in time. Sauntering in ancient towns filled with palaces and monuments, such as Hvar, Korčula and Vis, one requires very little imagination to reconstruct life as it must have been in these stone squares and terraces, in the loggias and streets, on the quays and in wine-cellars.

Linked for centuries with the centre of Hellas, the Greek settlements along the Adriatic gave out distant echoes of that great classical culture. Still enclosing the old towns, the walls of Hvar, Korčula and Vis are evidence of ancient architectural endeavours. Many museum exhibits found in Greek tombs on the islands show the elegant lines of Hellenic art. Take, for example, the gracefully draped *Tanagra*, or the figures on a terracotta vase fixed in an instant of dynamic movement. The craftsmen of ancient Pharos (Hvar), Issa (Vis), and Corkyra (Korčula) left works of exquisite refinement and skill.

Greek culture was supplanted by that of Rome which made its influence felt over a long period as it drew the whole coastland into its vast empire. Just how powerful the empire was, is evident in many places. On the islands the Romans erected magnificent buildings, architecture and setting in full harmony and displaying a fine artistic sense in the use of stonework. Roman stylistic devices, as elsewhere, had a lasting effect on the art of later centuries. When the Slavs reached the Balkans and the Adriatic, they built numerous small churches, cruciform in plan, pre-Romanesque in style. They bear witness to the Slavonic presence on the islands and coast and later became a symbol of Dalmatian architecture in miniature. The Romanesque influence continued several centuries on the islands. A splendid example of that period is the cathedral in Rab; its slender bell-tower stands separate from the church soaring skywards.

Stone remained the principal medium for artistic creativity on the islands. Shaped according to Romanesque tenets in Rab and Pag, the stonework gracing the town squares and streets of Korčula and Hvar conforms to Gothic and Renaissance dictates. In the eyes of a visitor Hvar town radiates the charm of a medieval stage setting: the main square flanked by honey-coloured facades, the cathedral with its blend of late Renaissance and Baroque styles, houses with old stone portals and elaborate coats-of-arms, and the many bell-towers of churches and chapels – stone caught up in a sprightly dance beneath the warm southern sun.

Born on the island of Brač, Ivan Ivanišević became bishop of Hvar in 1641, but fifteen years later he decided to forego the honour and devote himself exclusively to literature and historical study.

Besides cathedrals and churches, some dating back to the earliest Christian times, the islands were favourite locations for princely and aristocratic summer residences. The summer villa of Hanibal Lucić, the first Croatian dramatist, just outside Hvar town, reflects an engaging simplicity which says much about the author of one of the most charming love lyrics in early Croatian literature. Writer Petar Hektorović spent his summers in the small town of Starigrad on Hvar; his fortified villa, know as the Tvrdalj, has a plain exterior but encloses a courtyard and a delightful fish pond with here and there a line or two from the poet's verses carved into a stone lintel or threshold.

Most of these medieval buildings were the work of native stone-cutters and masons, each teaching his craft to the next generation and thereby ensuring precious artistic continuity. The best known of these craftsmen came from the Andrijić, Karlić, Macanović and Katičić families who worked on the cathedral in Dubrovnik and numerous other churches.

In interlocking patterns of stone the town clock counted off the hours till nightfall when lonely footsteps would echo through streets once thronged with sea-captains and merchants, playwrights and poets, for in the town's most prosperous period material wealth was matched by that of the mind and spirit. Visiting its churches and palaces I paused before the gateway of the Franciscan monastery. Passing through the cloister I came to the refectory. A whispered prayer *Ave Maria gratia plaena...*, the rustle of long robes, the faint click of beads. An older monk closing his prayer book; a younger one drawing water from the well. The sound of choir music... *Dies irae....*, wherever the wrath of the Inquisition led a militant church, the Franciscans marched as one body, the Benedictines as another. The past garbed in monks' cowls. A collection of embroideries, illuminated manuscripts, documents; shelves lined with ancient books; maps showing the Caspian Sea as charted by navigators in 1525. The monastery is full of valuable paintings: *The Glorification of St. Anthony of Padua* with a panorama of Hvar and the Pakleni isles. Marco Liberi, Jacopo Palma the Younger and a whole group of Renaissance painters trained in Venice. And dominating the room, is a six-metre-wide *Last Supper* painted by a grateful Matteo Rosselli. This was in the sixteenth century, a golden age for art in nearby Dubrovnik where the Venetian Rosselli had been spending some time. On the voyage home the painter fell ill, and his friends and the other passengers were afraid he had contracted the dreaded plague. Somewhere near Hvar the unfortunate artist was put off the ship in a rowing boat, abandoned to his fate. He was half-dead by the time the monks found him and carried him to the monastery. They nursed him for a long time and their care was such that the artist recovered. To show his gratitude he painted a magnificent *Last Supper* on the very spot it stands today. As models for the apostles he used sailors, fishermen and local peasants. But nobody wanted to be Judas. The monks suggested that the artist should paint from memory the face of the Venetian captain who had abandoned him at sea. The painter's own self-portrait is seen in the sickly servant gathering crumbs from the table to give to a dog.

In this picture we can see a detail from St. John and St. Simon *by Celestin Medović (1857-1920). This altar painting hangs in a parish church on the island of Pašman; its foundations were built in the early Middle Ages and the bell-tower in 1750.*

On a hill in the middle of the small town of Vrboska stands a fortified church. Serving two purposes, worship and defence, its walls have echoed gunshot from the altar, blessings after victory, the wail of a fisherman's wife just become a widow. Near St. Mary's is the parish church, which is dedicated to St. Lawrence. Its Baroque interior shelters a main altar polyptych by the Venetian artist Paolo Veronese, paintings by Bassano, Celestin Medović and Titian: *The Adoration of the Magi*, *The Birth of the Virgin*, *The Resurrection*, indistinct images in the shadowy play of light. Vrboska, Hvar, Jelsa, Starigrad. The patina bestowed by the past speaks of a constant swing of the pendulum, of illusion and ignorance, genius and mediocrity, flickering candlelight and streaks of lightning. In its dialogue with stone, memory sinks into the obscurity of the past leaving enigmas to stir the imagination. Throughout time man has been a witness; whether wearing monks' cowls or fishermen's boots, in prayer or profanity, he is the parchment of life.

In a Renaissance painting by Jacopo Palma, *The Virgin with Saints* in St. Stephen's Cathedral in Hvar, I picked out a face with the gravity of a Hvar peasant. The chasuble didn't suit him; neither did the bishop's scepter. His heavenward gaze was more convincing; he looked as if he were praying for rain. I found his twin in the village of Brusja on a hilltop overlooking Hvar. He was mending his nets in the barrel-vaulted shade of a stone house. On the gleaming white wall of the opposite house someone had written Street of Happy Love. Marko told me about his five daughters and how they had been courted. All the Romeos who had stood beneath his balcony, the serenades and sighs... Jelka, Dinka, Milica, Rosanda, Katica...

His wife came out with a pitcher of wine, a hunk of

cheese and a few olives glistening in the sun. With the olives she offered a smile, her movements restrained, her step as cautious as a deer, her toils reflected in a prematurely aged face. When she talked about her daughters, she beamed with happiness thinking about the future each had embarked upon. One was married to a florist, another to a lorry-driver, the third to a farmer. One smelled flowers, the second petrol and the third the earth... And one of the enamored youths had given the street its new name. Marko Pavlović had a fine vineyard which filled his cellar with wine, more than he knew what to do with. He had to go all the way to Mostar to try to sell it. The remainder turned to vinegar and was sold in that form.

The kitchen smelled of lavender and rosemary — heavy scents used in making French perfumes. The island fields are blue with lavender. Swinging a scythe, Marko cut through a blue tuft. His stone-walled field rose on a gentle slope behind the village. Wine may turn to vinegar, but lavender cooked, distilled and bottled can wait. For a litre of lavender oil one gets as much as for a hundred litres of wine. Labour seems immaterial in this scheme of things. A litre of lavender oil is the equivalent of a sack of blue flowers. And a sack of blue flowers — one day's work. From sunrise to sunset. In the old square of Hvar town, on the quays, at the market, peasant women set out tables with tiny bottles of home-distilled lavender oil and wait patiently for a customer.

In addition to a polyptych by Paolo Veronese and a Madonna of the Rosaries *by Leandro Bassano, the parish church of Vrboska, a village on the island of Hvar, also harbours this* Adoration of the Magi *by Celestin Medović (1857-1920) who studied painting in Rome and Munich.*

How lavender came to be grown on Hvar no one knows. A single plant blooms fifty years and perhaps longer. Its longevity and hardiness seem to reassure the islanders. Blue fields appear in the landscape backgrounds of Renaissance paintings. Did the artist have a predilection for blue, or are these lavender fields? Lavender was to be found on the shelves of the medieval apothecary in Dubrovnik's Franciscan monastery. Marko was still working on his nets, the sea was close by, and towards evening something might be caught.

People have lived on the island since time immemorial. In an abandoned hilltop cave, from which one has a panoramic view of the sea and an offshore island, Grga Novak, eminent Zagreb archeologist and native of Hvar, discovered traces of neolithic habitation, fragments of earthenware used by cave-dwellers. Later the cave had other tenants. In an age when the wandering Greeks were establishing their first island settlements, long before Pharos and Dimos were founded on the southern coast in 385 B.C., the caves were inhabited by a race of giants.

Just opposite the island of Šćedro is one of the largest groups of caves in the Mediterranean. The first time I came here it was with Aristid Vučetić, descendent of an old Hvar seafaring family. He died at the age of ninety-five, having devoted his whole life to researching an unusual thesis: that the wanderings of Homer's legendary hero had actually taken place in the Adriatic, that his ships returning from Troy had been driven by a tremendous gale through the Straits of Otranto and so into an unknown sea. And in one of the biggest caves on Hvar, the Grapčeva cave where Professor Novak discovered a neolithic site, lived the giant Cyclops Polyphemus. To the earliest settlers from Hellas, Hvar was known as Pitieia, and interestingly enough near the cave there is a small village called Pitve – perhaps a Slavonic variation of Pitieia. From the nearby island of Šćedro, the island with innumerable goats, Ulysses' sailors could hear the voices of the giants calling to one another from the many caves of Hvar. With a dozen men Ulysses crossed over to the opposite island where he had his terrible

Besides faithfully representing the churches and palaces, this 1646 print of Hvar town clearly shows the defence walls which were begun in 1278 and continued until 1806 when Napoleon's soldiers made further additions.

encounter with the Cyclops whom he finally managed to outwit, and with his surviving men escape from the cave. And today in the obscurity of the cave it is as if we could hear Ulysses himself...

"We climbed a steep path a hundred paces or so. The island was large and we were unable to investigate all of it. Tales reaching civilised Hellas relate of an island of Cyclopes who do not fear the gods, who never lift a hand to plant, all their crops springing up unsown and untilled. The grapevine bears its clusters without cultivation. Their strength is colossal; they eat whatever meat they find, whether mutton or human. It is said that the strongest among them is Polyphemus, son of Poseidon, the sea-god..."

Other links with the greatest wanderer of all time will be found on islands further south: Korčula, Mljet, and the Elaphite isles.

From time to time Ulysses returns to the green woods of Hvar. Present-day reality never quite shakes off the shadows of legend, the visions of the Venetian Rosselli, or Celestin Medović, artist from Dalmatia. Still present, to the musical accompaniment of a *leut* or *lijerica*, are the melodious verses of Lucić's love lyrics, the rhymed dialogue of Petar Hektorović's common folk, the fishermen. Hvar lives with the flickering past, as if the sunlight skipping over the listless surface of the sea were playing with time, and at every step we sense the colour and flavor of past events.

Our string of donkeys was filing slowly down a rocky hill to Jelsa, where we had begun our trek to the Cyclops' cave. The setting sun had abandoned our caravan to the pleasant twilight of a long summer evening.

A burst of laughter rang out. Beneath us lay the bay and Jelsa, a group of stone houses, a bell-tower and pine woods by the sea. The laughter of innocent tourists called up a growling Cyclops and the shrieks of helpless trapped men. Ulysses saw Polyphemus' jaws close down upon men with whom he had shared good and ill fortune since the Trojan War. The afterglow of legend touches our senses, and a visit to the Grapčeva cave is a trip into the past, into a world of epic poetry.

BRAČ

153. *Zlatni Rat beach near Bol on the island of Brač is one of the loveliest sandy beaches in the Adriatic.*

154. *Sunlight reflected on the sea in the channel separating Brač and Hvar.*

155. *Olive groves are the symbol of the Mediterranean. They grow very slowly and require great care if they are to bear fruit.*

156. Above. *The stone quarries on the island of Brač have had a lasting effect on island life. Its quality and durability of world-wide repute, this white stone has for centuries been the island's main export item.*

Below. *White stone from Brač, never tarnishing even in the worst atmospheric conditions, has been used to build many places, such fabulous triumphs of architecture as Diocletian's Palace in Split.*

157. *White Brač stone was used for the mausoleum of the Petrinović family in Supetar, a small town on the island of Brač. It is the work of a well-known Yugoslav sculptor, Toma Rosandić.*

158. Above, left. *Blaca, a cove west of Bol on the island of Brač.*

Above, right. *A peasant from the island of Brač. Besides fishing the islanders also cultivate vineyards which produce good wine.*

Below, left. *Conversation on a stone wall in the shade of a gnarled olive tree.*

Below, right. *In the scorching summer heat sheep seek shade in the thick scrub and wait for the sun to set. At dusk they amble to the watering-hole.*

159. *The rocky interior of the island of Brač dips into numerous valleys where the mild climate favours the cultivation of subtropical fruit.*

160. *A scene on the island of Brač: Glavice point with the Dominican monastery and church, founded in 1475.*

161. *The harbour of Bol, a small town on Brač.*

MARKO POLO

9 10 11 12 1 2 3

KORČULA
A BATTLE WON BY THE WIND

KORČULA

162. *The roofs of ancient Korčula. A close-packed ensemble of houses circled by walls. A patchwork of colourful, irregular pieces tells stories about the people who live, and used to live beneath these roofs.*

163. Above. *This late Baroque staircase with its decorative balustrade was built in 1548 and leads from an open loggia up to one of Korčula's main gates.*

Below. *One of the narrow streets of old Korčula which lead to the main square and the cathedral.*

164-165. *A view of Korčula town from the air. This is one of the most beautiful and best preserved of the ancient towns on the Adriatic islands. Today it has a population of 2,500 and welcomes growing numbers of tourists. Straddling a promontory which juts out into the Pelješac channel, the old town is encircled by walls built over a period running from the 14th to the 16th centuries.*

166. Above. *The house where the celebrated traveller Marco Polo is supposed to have been born. His memoirs describe his adventures travelling through the Far East.*

Below. *The remains of an ancient wall near Lumbarda, a village on the island of Korčula. It is believed that this was the site of the goddess Circe's palace where Homer's legendary hero Ulysses was entertained during his wanderings in the Adriatic Sea.*

167. *A sundial on the facade of an old house in Lumbarda. In these sun-drenched regions time was told by shadows which marked the sun's path across the sky.*

168. *Between vineyards and the seashore near Lumbarda there are the ruins of a manor house of more recent date.*

A narrow channel separates Korčula from the peninsula of Pelješac. From this distance the island and its ancient fortified city looked dark and hazy, only flecked with sunlight. In Greek times it was known as Korkyra Melaina, and ancient mariners told many tales about it.

Near the village of Lumbarda we found strange-looking ruins, dressed stone blocks positioned at a slant. Only the foundations remained; the rest had been hauled off by peasants to make walls for their vineyards. If one pieces together legend, ruins, island names, as well as research which has traced the wanderings of Ulysses to this spot, the enchantress Circe had her palace here in the fields of Knežina — near a medieval chapel and vineyards which produce a wine called, oddly enough, Grk (Greek). Today vacationers pass between the tall, leafy rows of grapevines to reach the beach at the cove of Pržina where it is believed the great Ulysses accepted the favours of the goddess Circe. On the smooth rocks sloping down into the emerald seabed Ulysses sat longing for his native Ithaca and the faithful Penelope.

Thousands of years have passed, yet the message of Homer's epic still rings true. Longing for a distant homeland, for one's dearest and the warmth of a hearth is as universal and timeless as pain, wanderlust and war. That which is human, whether good or evil, endures and is repeated throughout time. Some attack, others must defend themselves, still others wander and search. Korčula's story goes back a long way. After the Greeks and the Romans and the Slavs from the Neretva valley who were the first settlers, the island fell into various hands. People huddled together behind walls in fear of attack and epidemic; the less turbulent periods produced a flowering of the arts and crafts. History repeats itself on the Dalmatian islands.

The islands survived their many conquerors and the walls girding the medieval towns testify to the resilience of the freedom-loving inhabitants. In a web of stone streets and squares, Korčula, too, has a similar history of events.

In 1571 Korčula's population of six thousand was greatly depleted by the plague. Many houses were burnt to keep the deadly disease from spreading, their ruined walls are still standing. The townspeople defended themselves from the plague with fire; from attack with high walls and forts. As legend has it, the city was founded by the hero Antenor who lived several centuries before Christ. Inside a small tower near the twin obelisks raised in honour of governors Telani and Pasqualigo is a stone slab dated 1592 evoking the memory of Antenor.

The town was entered through three gates with staircases. The one at the far end, the Land Gate, served as an exit in time of danger through which the population could withdraw to safety in the rocky interior of the island. The second, wider and more monumental, opened to welcome Dubrovnik princes, bishops and processions. In the evenings until the lamplighter made his last rounds, young lovers used to meet on the stone steps, youths came to recite poetry, or sing some newly fashionable song.

The Renaissance-Baroque palace adjoining the cathedral, once the residence of Korčula bishops, now houses the Abbey treasury with a collection of paintings, sculpture and precious objects from local churches. This picture is part of a triptych dated 1431 by Blaž Trogiranin.

On numerous occasions the town walls proved too strong for the enemy. One battle, however, passed into legend: in 1571, after sacking several other Dalmatian towns, the Turkish renegade and pirate Uluz-ali attacked Korčula. Having slipped through a breach at Lepanto, Uluz-ali sailed with his galleys into the Adriatic. When the Venetian governor, the nobility and the clergy of Korčula heard that a Turkish fleet was approaching, they packed up and fled in panic. Only the common folk, the women and children were left behind. Resisting as best they could, they had just about given up all hope when a strong north wind blew up snapping the anchor chains of the Turkish galleys. The women and children who were on the ramparts manning the guns could see the enemy ships withdrawing. There were few casualties and no one had been taken captive.

Although tales about the battle with Uluz-ali are still told, people in every home like to tell about a former inhabitant of the island who travelled far and wide. Near the cathedral is the house of the celebrated traveller and counsellor of kings Marco Polo, whose memoirs describe life in the court of Kubla-Khan on the Mongolian steppes, his adventures in China and India and the customs of Tibet.

St. Mark's Cathedral on the main square, Gothic with elements of later styles, was completed towards the close of the fifteenth century. With so much stone at hand it is not surprising that the islanders became such skilled stone-cutters. Work on cathedrals generally occupied several generations, and each hoped its own contribution would surpass in beauty all that preceded. As styles changed from Romanesque to Gothic, to Renaissance, the stone-cutters of Korčula succeeded one another, leaving works which show their imagination, religious faith and industry. Above the delicate rose-window rises a slim bell-tower. Connected with this stonework and sculpture are the names of men who shaped stone into an expression of their own concepts of the world: Dragošević, Ivančić, Hranić, and Marko Andrijić who executed the ciborium which adorns the main altar together with an early Tintoretto depicting St. Mark with Saints Bartholomew and Jerome. Times, ideas and people change; the carved stone figures, like legend, survive intact.

The people of Korčula treasure their traditions. Every year on July 27 costumed dancers assemble on the main square before the cathedral to commemorate the heroic defense of the town five centuries ago. This is St. Theodore's Day, the day of the Moreška sword dance. Two groups of dancers, one in black, and the other in red costumes, take up positions. Each group has a leader wearing a golden crown. The dancers in black represent Turkish soldiers; their leader is called Osman. The red-clad dancers are the Moors; their leader is Moro. The fight is over a young captive veiled in white, whom one of the Moors holds prisoner. Each selects his opponent and the swords cross, bodies bending with the swinging swords. The tempo increases as the clashing swords whirl faster and faster. These are not, however, duels, but an ensemble in which each participant tries to maintain a perfect coordination of movement in an accelerating rhythm. Finally after a long, exhausting contest, one side admits defeat laying down its arms, and withdraws. The victors free the captive and celebrate with a quick-paced dance which represents the climax of a pageant of unknown origin. It is believed that sailors saw dances with similar themes in the Mediterranean and brought them back to the island.

St. Mark's Cathedral in Korčula (begun in the 13th century and completed in the 16th) harbours an Annunciation *by Tintoretto's school. The Virgin Mary is shown here.*

The Elaphite Isles

ŠIPAN

Olive groves, vineyards and fig trees

We were sailing through a narrow passage between the Pelješac peninsula and the small island of Jakljan, our steamer making slow headway against the strong currents. Set squarely in the middle, the isle of Olipa divides the narrows into two channels: to the left Scylla, to the right Charybdis. Rolling breakers thunder against the sheer cliffs, growling like the six-headed Scylla. In a black cliff to the right the gaping mouth of a cave sucks in the heaving sea: the insatiable Charybdis. We were again on the path of Ulysses and the groaning waves stirred the imagination. Once through the narrows and in the soothing, calm waters of the Elaphite archipelago, we began looking out for the entrance to Šipan's harbour. A string of verdant isles offer sandy beaches tucked away inside tempting green-fringed coves. Jakljan, Šipan, Lopud, Koločep, Lokrum — the last offshoots of the Adriatic islands. Officially they are Dubrovnik's islands. Pliny the Elder called them *Insulae Elaphitae* (deer islands). They were favourites with the Romans, welcome refuge for pirates from the Neretva and Ulcinj, excursion spots and summer retreats for the Dubrovnik nobility. Ancient legend links them with the islands where the Sun-god Hyperion kept his cattle and sheep. Today on the green island pastures in wintertime one sees flocks which herdsmen ferry over from the mainland mountains.

Passing between Mišnjak point and the isle of Crkvine we entered the harbour of Šipan: an island of olive groves, vineyards, carob and fig trees, almonds, pomegranates and oranges. Rising out of the luxuriant vegetation are the ruins of old churches and Renaissance palaces. One finds St. Peter's, St. John's, and St. Michael's on Šipan, another church dedicated to St. Michael on Lopud, and St. Nicholas' on Koločep. They were built in the eleventh century of roughly dressed stone with small blind arcades, miniature domes and bells that after nine centuries still ring. What do they remember, whom did they summon out for an afternoon stroll? In the fifteenth century the aristocratic Sorkočević family of Dubrovnik had a residence here. Above the harbour are the walls of the Dubrovnik prince's palace: Gothic windows, on the facade, and over the portal the date 1450 is inscribed. Further on in Sudjuradj stands a fortified castle built by Tomo Stjepović Skočibuha; his son Vice added a tall tower in 1577 as protection against pirate raids. The Elaphite islands are dotted with churches, castles and summer homes. Between the harbour and Sudjuradj are the ruins of the summer residence of the Dubrovnik archbishops which the Italian humanist Lodovico Beccadelli, a friend of Michelangelo's, visited in the sixteenth century. In Renatovo are the ruins of a smaller castle where legend claims that René of Anjou, King of Naples, once resided. His coat-of-arms, bearing the motto *Renatus rex justus*, was found in the ruins.

Round the palaces, summer homes and churches are fields with vineyards and thick grass. At one time the first hours of dusk were announced by the cries of jackals. The jackals have long since been killed off, only the song of the

The Republic of Dubrovnik, its great fleet rivalling that of Venice, had its own shipyards on the Adriatic islands. The picture shows a large ship of the Republic of Dubrovnik on the island of Šipan in 1806.

crickets remains. On these fields, despite the warning of the goddess Circe not to harm the Sun-god's cattle, Ulysses' starving and desperate men, after all the horrors of Scylla and Charybdis, rounded up and slaughtered several of the cows. Fearful of the consequences, the men sat gazing at the fires, and as it seemed to them the hides crawled, the meat groaned on the spit, a sound as though of lowing cattle. In fact what they saw and heard were jackals dragging off the hides into the night. As the story continues, Ulysses' ship ran into a gale between Jakljan and Šipan. A bolt of lightning struck the mast and the vessel sank. Only Ulysses was spared, drifting through the channel to Mljet and the embraces of the nymph Calypso.

The Argonauts with Jason also sailed through the narrow straits of the Elaphite islands in their search for the golden fleece. In the Harpoti narrows several centuries later a Roman fleet came to grief. Classical legend gives the name of Thrinacie to the three largest islands: Jakljan, Šipan and Lopud. A fresh water spring *šipun*, which never ran dry, gave the island of Šipan its present name. On the islands of Thrinacie two goddesses, Phaethusa and Lampetie, herded the Sun-god's cattle. On winter days the flocks belonging to peasants from Herzegovina and southern Bosnia complete the picture described in Homer's epic.

LOKRUM
THE NATIONAL PARK

Southernmost in the chain of Elaphite isles, just opposite the walls of the city of Dubrovnik, lies forest-clad Lokrum. Its name is probably derived from the Latin *acrumen* (bitter fruit). The rare flora that abounds on Lokrum has caused it to be declared a national park. The exotic plants brought home by sailors over a period of several centuries turned the island into a lovely garden. In 1023 the Benedictines built a monastery and church; later additions left the bell-tower, which also served for defence, inside the triple-aisled Gothic church. The church, monastery and parts of the walls fell in the disastrous earthquake which levelled Dubrovnik and its surroundings in 1667. The southern coast of the island is steep and inaccessible. The sea roars thunderously into deep clefts, and at nightime, when the sea is calm, night-swallows swoop down, their cries echoing from the rocky hollows. The whole island is covered with tall-growing scrub, thick with vines, and on the southern part dark woods of Aleppo pine. The scent of sage and wormwood hangs over scattered green fields. It was on this island in 1192 that the English king Richard the Lionhearted, returning from the Crusades, sought shelter during a fierce storm. At the time of the Napoleonic wars the French built a fort on the highest point of the island: Fort Royal, its bastions spreading out into a five-pointed star. In the 1830's Austria added the Maximilian tower to the French fort. Lokrum was purchased by the Hapsburg Archduke Maximilian in 1867, and in the ruined part of the Benedictine monastery he had a palace built according to his own plans, like the castle of Miramare in Trieste. After Maximilian's execution in Mexico the castle had several owners including the Austrian Crown Prince Rudolph.

Miho Pracat (1522-1607), who owned and captained a number of sailing vessels, was one of the wealthiest and most influential citizens of Lopud. His ruined summer villa still stands on the shores of his native island. (Left)

In Sudjuradj, a small village on the island of Šipan, there is a mansion which Tomo Stjepović-Skočibuha had built in 1539, while his son Vice (in the picture), seaman and ship-owner, added a tall tower for defence. (Right)

Present-day strollers along the narrow, flower-banked paths of Lokrum are more likely to be tourists, yet the island doesn't seem to mind. The winter winds and spring rains bring respite from the incursions. During these periods the only visitors are biologists studying the flora and fauna of the Adriatic coastland. Within the old monastery walls is the Biology Institute of the Yugoslav Academy of Arts and Sciences of Zagreb.

The Final Notes of a Symphony

From the great gulf of the Adriatic the sea thrusts a strong watery fist deep into the rocky mountain barrier of the coastland – the Bay of Kotor, or in the words of a Yugoslav poet "the bride of the Adriatic." In this magnificent corner of the southern Adriatic the rugged mountains are mirrored in the ever smooth surface of the sea, and one can easily picture the famous sailors of Kotor rigging their ships for long ocean voyages.

In the middle of the bay, where it branches out into the smaller twin bays of Risan and Kotor, lie two islets: Gospa od Škrpjela and Sveti Djordje, like two dots marking the end of a long, unfinished sentence, each word standing for one of the islands of the Adriatic. On tiny Sveti Djordje, where the Benedictines built a monastery in 1166, one can see walls, cypresses, a tower and the red roof of the church. A dozen metres beyond is another islet which owes its existence to the seamen of Perast, its weathered stone facades reflected in the sea not far from the island. Gospa od Škrpjela did not rise from the sea in some early geological era. It was built. In the seventeenth century, after a Turkish fleet had been defeated by galleys from the Bay of Kotor, the captured vessels were filled with rocks and sunk near the isle of Sveti Djordje. Whenever the sailors returned from a voyage they added a few more rocks to the growing pile. Eventually an islet was formed, and in 1630 a church was built, called Gospa od Škrpjela. Both Italian and native artists contributed to the decoration of the interior. The figures of the saints are the work of the Venetian sculptor Francesco Gaia; the painting of the Virgin on the high altar was executed in the fifteenth century by a native artist, Lovro Marinov Dobričević. The walls and ceiling of the tiny church, its foundations resting on Turkish galleys, are covered with 68 paintings on canvas. The Old Testament

GOSPA OD ŠKRPJELA AND SVETI DJORDJE

TWO ISLETS IN THE BAY OF KOTOR

MLJET

177. *This small cove is known for its oyster beds. Several restaurants nearby have oysters on the menu.*

178-179. *Verdant Mljet, one of the loveliest islands in the Adriatic, has been made a national park. Its flora and fauna will be preserved for future generations to admire.*

ELAPHITE ISLES

180. *This wooded islet has had an eventful past. In 1023 Benedictine monks settled on Lokrum, and since then there has been a succession of masters and visitors from Richard the Lion-Hearted to Emperor Maximilian and the Austrian Crown Prince Rudolph. The island is also a national park.*

181. *A view of the Elaphite isles – with Dubrovnik in the foreground – "deer islands" as Pliny the Younger called them. The three largest are Šipan, Koločep and Lopud.*

prophets and the scenes from the lives of Mary and Christ were painted by Trifun Kokoljić, a well-known Baroque artist from the Bay of Kotor. Between the two zones of paintings in the nave hang about two and a half thousand silver votive plaques with engravings that illustrate life in seafaring Perast.

Both islands had cannons trained on the entrances to the twin bays to protect them from frequent pirate raiding parties.

SVETI STEFAN
The Yugoslav Capri

The third dot at the end of our unfinished sentence lies off the sandy shores of the southernmost part of the coast. Minute Sveti Stefan, the only island in this part of the Adriatic, is not far from ancient walled Budva. One could say that the thousand and one islands of the Adriatic begin in the north with Sveti Stefan near Ankaran and end with another Sveti Stefan near Budva in the south. A stone causeway connects the islet with the mainland, sandy beaches fanning out in either direction. As legend relates, the island was settled by the Paštrović clan who fortified it on the land side and hurried there whenever danger threatened. Otherwise, the small stone houses were used for storage; stone olive presses and basins for the oil have been found among the ruins. The oldest building is the fifteenth-century Church of St. Stephen. Eventually the clan moved permanently to the island where they felt safer and began to build houses to live in. Centuries later, when the pirate raids ceased, Sveti Stefan was abandoned as its inhabitants made their homes on the mainland. The deserted village is now a hotel, the old houses having been converted into more than two hundred apartments.

The holiday-makers who stroll along its narrow streets are actually bringing the old town, now "the Yugoslav Capri," back to life.

Islets in the Bay of Kotor

182-183. *Two islands in the Bay of Kotor near the medieval town of Perast. On the one to the right, named after St. George, stands a Benedictine monastery built in 1166. The island to the left, Gospa od Škrpjela, is actually a pile of rocks, with the church built on top in 1630.*

184. *The "island of flowers", as it is called, in the Bay of Kotor near Tivat. A bridge spans the water to the mainland. An attractive holiday village nestles in luxuriant vegetation.*

185. *An old door with its typical stone frame. Architecture in these coastal communities has an attractive rustic simplicity. Already belonging to the past, houses such as these are giving way to modern apartment buildings.*

Sveti Stefan

186-187. *A strip of sand connects the southernmost island with the mainland. The first houses on medieval Sveti Stefan were used to store food supplies. It was subsequently fortified by the Paštović clan who moved there permanently. Today Sveti Stefan has been converted into a hotel.*

Black Caryatides

To conclude, in lieu of an epilogue I would like to add my recollections of a motif which is inevitably encountered in the Adriatic, and since it has a special meaning for me I am taking advantage of this opportunity to help future pilgrims to the Adriatic understand.

As I travelled from island to island I saw many black headscarves. On the island of Krk I was told that the women were mouring the Croatian prince Frankopan, medieval ruler of the island. On islands such as Lošinj, which had an ancient seafaring tradition, they explained that in the days of sailing vessels many seamen never returned. While the menfolk were absent, the women wore black scarves, and when they came home the women continued wearing the scarves for the men who had not yet made port. Thus the black scarves became a customary part of women's dress in the Adriatic. A kind of symbol.

In a small fishing village I visited thirty years ago I found only one man. He told me how the fishermen used to return every evening, dock at the quay and hand up crates filled with fish to the women. That was where life came from.

Then the war came.

All the fishermen took to the mountains and for four years fought for survival. Every evening the women went down to the waterfront and waited, hopefully and anxiously. The war ended and the long-awaited peace finally came. But none of the fishermen returned to the village. The women still went down to the shore in the evening, standing there for hours, their gaze turned seawards, black marble caryatides supporting the heavy burden of a widow's life.

Many years later I passed through the village again. Dotting the beach were the brightly-coloured tents of a holiday camp. And near the last house at the edge of the rocky wastes stood a stone tablet bearing the names of the fishermen who had lost their lives in the war. A flower had been placed on the stone. On the sand round the tents children were playing with a big ball, chirruping like birds. Round the point came fishing boats; they drew near the quay where the women stood waiting. They placed the crates of fish on their heads and carried them home. A short while later I saw campers walking into the village to buy fresh fish. The fishermen were very young; they were the children I had seen many years ago, right after the war, playing on the sand while their mothers waited for the fishing boats to sail into the cove.

Through a door where a young woman was nursing her child I noticed on a chest, next to an image of the Virgin, the photograph of a soldier yellow with age and the smoke of candles. On a lace cloth under the picture lay a medal. The baby was crying, pushing his little fists at his mother's breast. It was early evening and the black kerchiefs were already on the shore.

From the camp came the voice of a pop singer.

Everything else faded into silence. The sea heaved a sigh, quiet and deep...

AU REVOIR!

Now all you have to do is pick out one of the islands you have been reading about. There are many more whose tales have not been told, but they would be similar, almost identical, speaking of beauty, sunshine, the kindness of the people, their eventful past and traditions. So if you visit an island and decide you like it, call it *yours*. You can even carve your name in some secret spot on a smooth rock by the sea. Next to your name write *Au revoir*, because every time you return to your island it will be with the same pleasure, with the feeling that while bathing in the sun and the sea you have found yourself again.

The first printed books, incunabula, were brought to the Adriatic islands by priests and humanists educated in Italy. This illustrated page comes from a book printed in 1494 and now in the museum of the Dominican monastery near Bol, a small town on the island of Brač.

Tourist Information

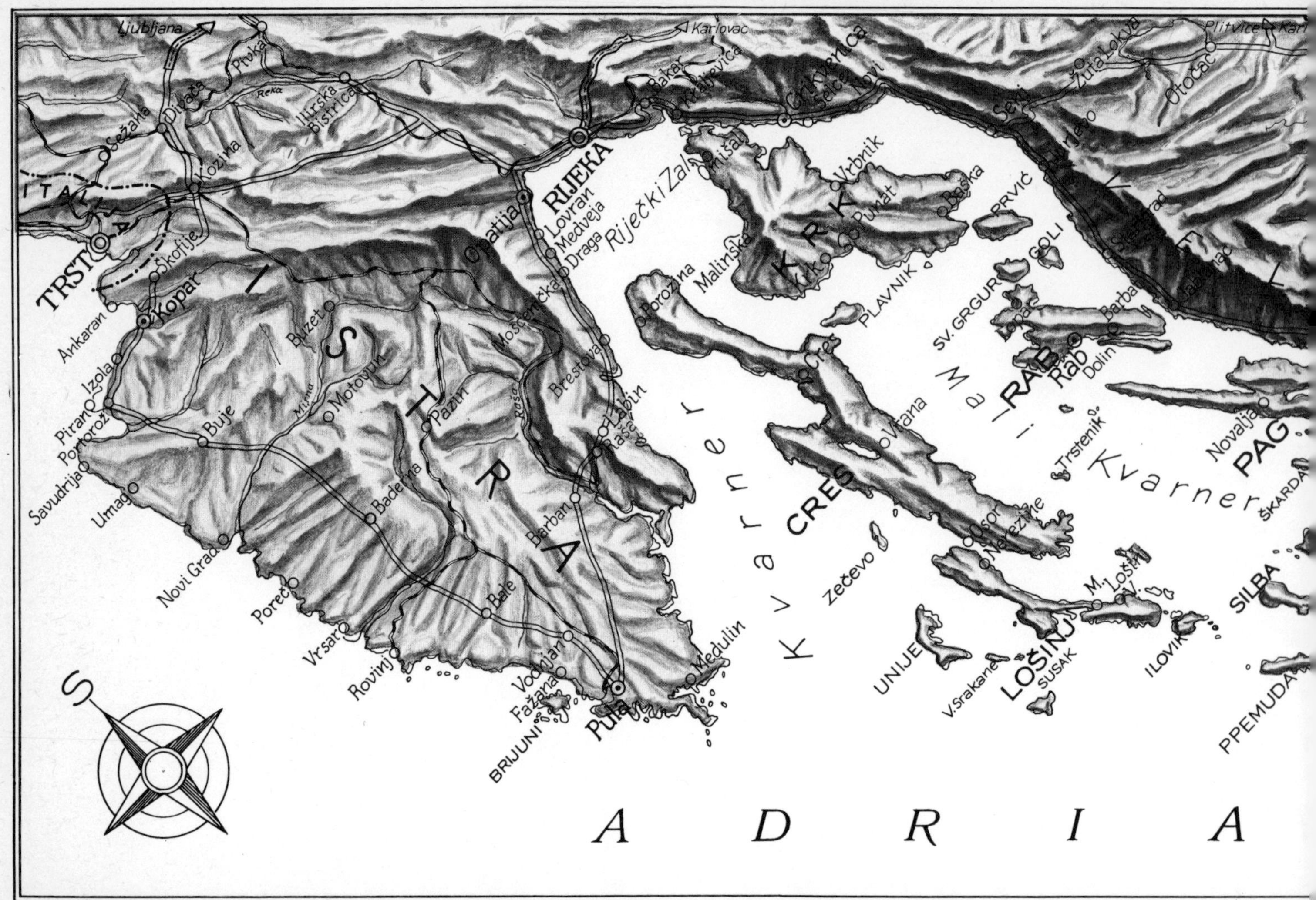
TRST
RIJEKA
ISTRA
KRK
CRES
RAB
PAG
LOŠINJ
Kvarner
Kvarnerić
Pula
Opatija
Kopar
Poreč
Rovinj
A D R I A

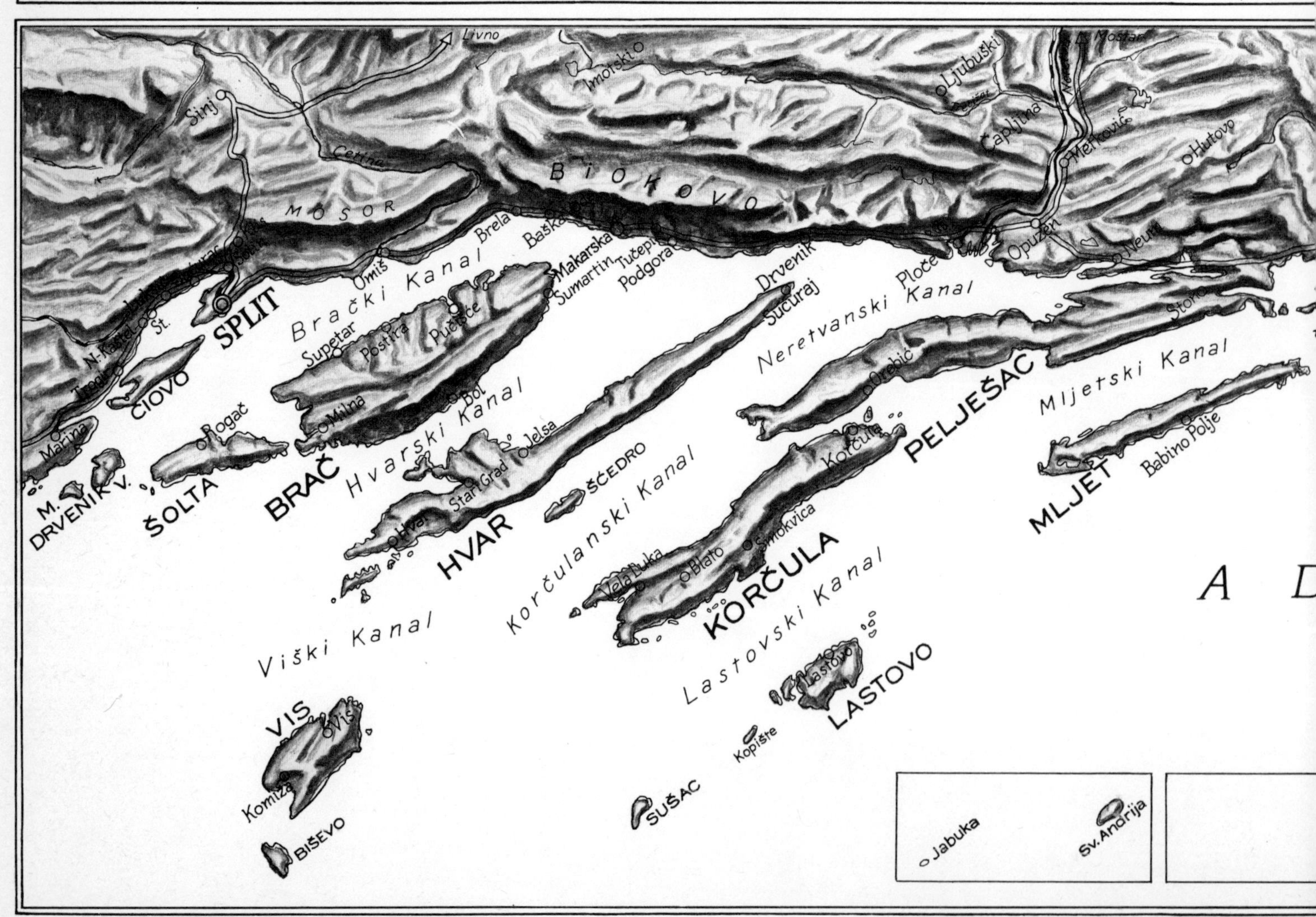
SPLIT
BRAČ
HVAR
ŠOLTA
KORČULA
PELJEŠAC
MLJET
VIS
LASTOVO
Brački Kanal
Hvarski Kanal
Korčulanski Kanal
Neretvanski Kanal
Mljetski Kanal
Lastovski Kanal
Viški Kanal
Makarska
Biokovo
MOSOR
A D

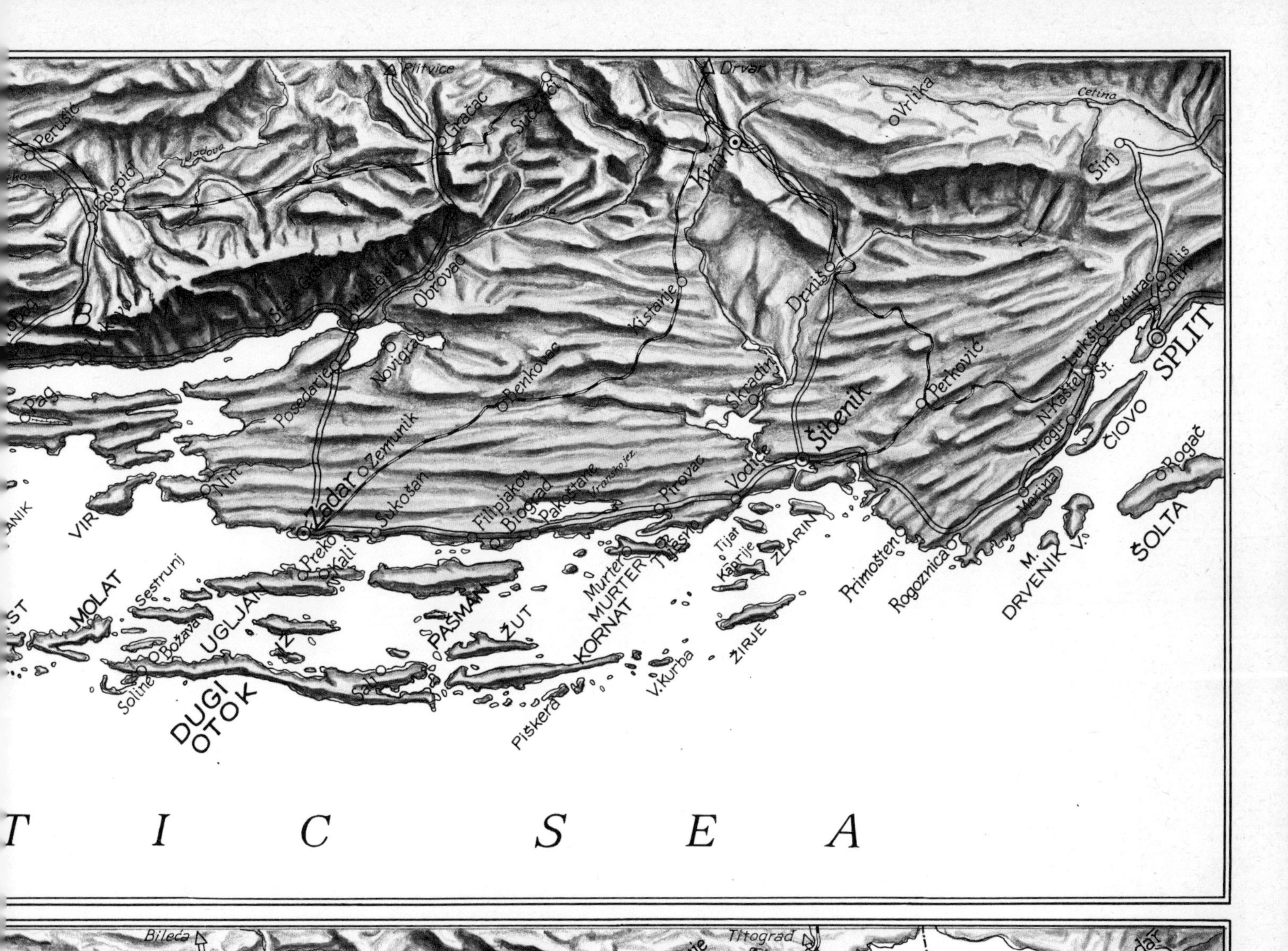
Plitvice
Drvar
Gračac
Knin
Sinj
Gospić
Obrovac
Kistanje
Drniš
Klis
Solin
SPLIT
Novigrad
Benkovac
Posedarje
Zemunik
Skradin
Šibenik
Perković
Trogir
ČIOVO
Pag
VIR
Nin
Zadar
Sukošan
Biograd
Pakoštane
Pirovac
Vodice
Primošten
Rogoznica
ŠOLTA
Rogač
MOLAT
Sestrunj
UGLJAN
Preko
PAŠMAN
ŽUT
KORNAT
MURTER
ZLARIN
ŽIRJE
Tijat
Kaprije
M.
DRVENIK V.
Božava
Soline
DUGI OTOK
Piškera
V.Kurba
T I C S E A

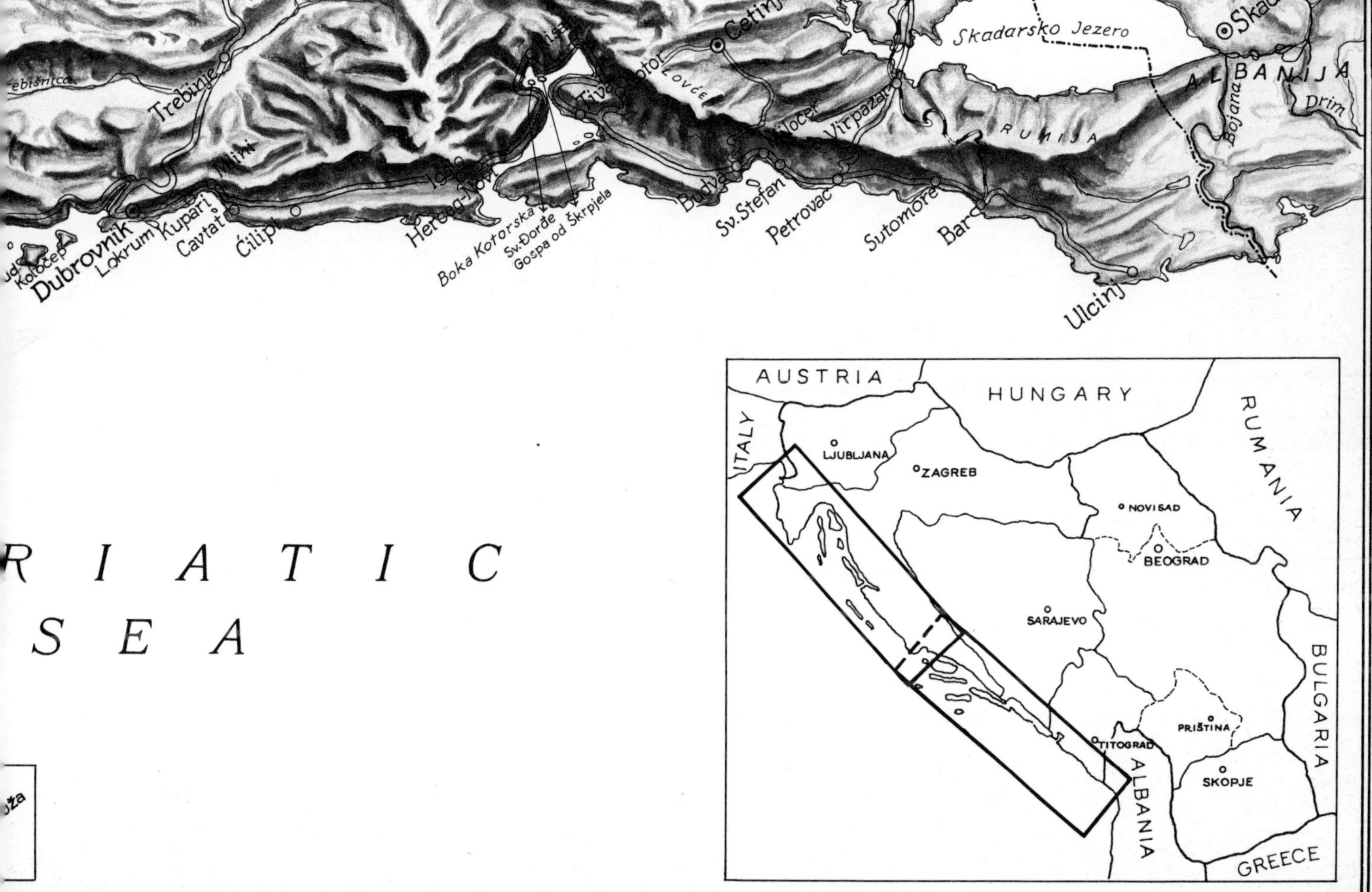
Bileća
Titograd
Trebinje
Cetinje
Skadarsko Jezero
Skadar
ALBANIJA
Drim
Bojana
Lovćen
Virpazar
Dubrovnik
Lokrum
Kupari
Cavtat
Čilipi
Herceg-Novi
Boka Kotorska
Sv.Đorđe
Gospa od Škrpjela
Budva
Sv.Stefan
Petrovac
Sutomore
Bar
Ulcinj
Koločep
R I A T I C
S E A
AUSTRIA
HUNGARY
RUMANIA
ITALY
LJUBLJANA
ZAGREB
NOVISAD
BEOGRAD
SARAJEVO
PRIŠTINA
TITOGRAD
SKOPJE
ALBANIA
BULGARIA
GREECE

Only a hundred years ago this schooner was proudly sailing the Adriatic. This was the Ida Maggiore, *a brig based in Koločep in 1885.*

CLIMATE AND SEA

The Adriatic islands have an average of 2,600 hours of sunshine annually (Palermo in Sicily, for instance, has 2,300 hours); they are thus comparable with Andalusia in Spain, the sunniest region of the Mediterranean. The sea temperature around the Adriatic islands approaches 14° C in February, reaches 24° C in July and is still 20.5° C in October, making an annual average of over 18° C.
The average air temperature during the spring months ranges from 13.5° to 16° C; in summer it reaches 24° to 25° C (not uncomfortably hot), while in the autumn it is extremely pleasant with an average of 16.4° to 17.7° C. During winter the mercury rarely falls below 10° C.

HOW TO GET THERE

By air

Yugoslav and foreign airlines and charter companies use the following airports along the Adriatic coast: Pula (for Istria), Rijeka (on Krk island) for the Kvarner and Istrian regions; Zadar (northern Dalmatia), Split (central Dalmatia), Dubrovnik (for the Dubrovnik Riviera, Herceg-Novi and the Montenegrin coast), and Tivat (closest to the far south of the Yugoslav coast).
Air connections can also be made through other Yugoslav airports: Ljubljana, Zagreb, Belgrade, Skopje, Sarajevo and Ohrid.

By road

The majority of motorists coming from or via Switzerland and Italy take the Milan-Venice-Trieste motorway, entering Yugoslavia by one of the frontiers in the Trieste region.
An attractive route across the Dolomites runs via Tarvisio and Predil to Bovec and the Soča valley, and thence to Sežana and the northern Adriatic.
The best crossing point from Austria is Potkoren, in the Wurzenpass. The route runs via Ljubljana to the Slovenian coast, Istria and the Kvarner region.
From Hungary, the quickest way to reach the Adriatic coast is via Subotica, Novi Sad, Ruma, Šabac, Zvornik and Sarajevo.
Many other combinations for reaching the sea can be worked out, depending on the motorist's frontier crossing point.

BY BOAT AND FERRY IN THE TOURIST SEASON

In summer, regular shipping lines operate from Pula (Istria) calling at some of the Adriatic islands (Lošinj, Hvar, Korčula). Ferry-boats connect the Italian and Yugoslav shores of the Adriatic (from the Italian ports of Ancona, Bari and Pescara to the Yugoslav ports of Zadar, Dubrovnik, Split and Bar).
Local ferries connect the mainland with the islands of Cres and Lošinj (from Rijeka, Brestova and Pula), Krk (from Crišnjeva near Kraljevica, Crkvenica and Senj), Rab (from Senj, Lopar on Krk, and Jablanac), Pag (from Jablanac and Karlobag), Ugljan (from Zadar), Pašman (from Biograd), Šolta (from Split), Brač (from Split and Makarska), Hvar (from Split and Drvenik) and Korčula (from Orebić).

By rail

Railway services from all parts of Europe provide connections (more frequent during the holiday season) with the following places on the Adriatic: Rijeka (from Vienna via Ljubljana, from Bucharest via Zagreb, etc.), Split (from Stuttgart via Zagreb, from Budapest via Zagreb, etc.), Ploče (from Munich via Zagreb, from Vienna via Zidani Most, etc.), Bar (from Hamburg, Stockholm, Basel via Belgrade).

BRAČ

The third largest island of the Adriatic. Towns and villages: Bol, Milna, Postira, Povja, Sumartin, Supetar, Sutivan, Pučišće, Bobovišće, Splitska, Selce.
Accomodation: in hotels and private homes.
Sports facilities, recreation, entertainment: marinas, camping, tennis, mini-golf, bowling, pools, naturist beach, water-skiing, underwater fishing.
Boat services: ferryboat lines Split-Supetar, Makarska-Sumartin

CRES

Towns and villages: Cres, Martinišćica, Punta Križa, Valun, Porozina, Vrana.
Accommodation: in a hotel and private homes. Campsites.
Naturist campsite and beach; marina, sailing, fishing.
Boat services: Cres is connected by a bridge with Lošinj, so that the boat services are the same as for Lošinj: ferryboat lines Rijeka-Porozina, Brestova (31 Km. southwest of Opatija)-Porozina, and Pula-Lošinj-Zadar.

HVAR

Towns and villages: Hvar, Stari Grad, Vrboska, Milna, Jelsa, Sućuraj, Zavala, Dubovica, Nedija.
Accommodation: in hotels and private homes. Campsites.
Sports facilities, recreation, entertainment: marinas, underwater fishing, tennis, bowling, mini-golf, water-skiing, naturist beaches.
Boat services: ferryboat lines Split-Vira, Split-Stari Grad, Drvenik-Sućuraj, steamer lines from Split, Dubrovnik, Rijeka.

KOLOČEP

Villages: Gornje Čelo, Donje Čelo.
Accommodation: in a hotel and private homes. A campsite.
Sports facilities, recreation, entertainment: sailing, field sports.
Boat services: small steamers from Dubrovnik.

KORČULA

Towns and villages: Korčula, Vela Luka, Lumbarda, Blato, Brna, Čara, Prigradica, Prižba, Smokvica, Raćišće.
Accommodation: in hotels and private homes. Campsites accommodating several thousand persons and naturist holiday village.
Sports facilities, recreation, entertainment: marinas, light and heavy repairwork on sailing craft, indoor and outdoor pools, casino, fishing, sailing, underwater fishing, water-skiing, tennis, bowling, mini-golf.
Boat services: ferryboat lines Orebić-Korčula; to avoid driving the length of Pelješac peninsula one can take a ferry from Ploča on the mainland to Trpanj on the northern coast of Pelješac; steamer lines from Split and Dubrovnik, with airports nearby.

KORNATI ISLES

A group of 125 islands of which only one is inhabited.
Accomodation: a limited number of fishermen's cottages on certain islands can be rented (further information may be obtained in Zadar).
Boat services: transportation in fishing boats or sports craft can be arranged. The main port of access is Zadar, although it is most convenient to visit the Kornati isles if one is already based on one of the islands belonging to the Zadar archipelago: Silba, Olib, Premuda, Iž Veli, Ugljan (villages: Ugljan, Preko, Kali, Kukljuca), Pašman, Dugi Otok, where one can find accommodation in hotels and private homes; campsites with a capacity of several thousand persons.

KRAPANJ

Accommodation: in private homes.
Water sports and underwater fishing.
Boat services: the main port of access is Šibenik; Krapanj and Zlarin belong to the Šibenik archipelago of some thirty islands which are served by small steamers; information about their schedules can be obtained at the actual place of departure.

KRK

The largest of the Adriatic islands.
Towns and villages: Krk, Baška, Malinska, Njivice, Omišalj, Punat, Glavotok, Šilo, Vrbnik.
Accommodation: in hotels and private homes. Campsites. Naturist beaches.
Sports facilities, recreation, entertainment: marinas, light repairwork on yachts and other craft, underwater fishing, tennis, bowling, mini-golf, water-skiing, sailing, casino, indoor and outdoor pools, rent-a-car service.
Boat services: ferryboat lines Črišnjeva (near Kraljevica)-Voz, Crikvenica-Šilo, Senj-Baška.

LASTOVO

The places on this island are not yet equipped to accommodate tourists, and information about day trips can be obtained in travel agencies on the mainland or in larger towns on the islands.

LOPUD

Village: Lopud.
Accommodation: in hotels and in private homes. A campsite.
Sports facilities, recreation, entertainment: sailing, water-skiing.

LOŠINJ

Towns and villages: Mali Lošinj, Veliki Lošinj.
Accommodation: in hotels and private homes. Several thousand

persons can be accommodated on campsites. Naturist campsites and beaches.
Sports facilities, recreation, entertainment: marina, light repairwork on boats, indoor pool, sailing, water-skiing, bowling, mini-golf, underwater fishing.
Boat services: ferryboat lines Rijeka-Porozina, Brestova-Porozina, and Pula-Lošinj-Zadar.

MLJET

Villages: Govedjari, Polače.
Accommodation: in a hotel and in private homes.
Sports facilities, recreation: pleasure craft may be anchored independently, underwater fishing, beaches in unspoiled natural surroundings.
Boat services: regular steamer lines and daily excursion boats, generally from Dubrovnik.

MURTER

Village: Murter.
Accommodation: in a hotel and in private homes. Campsites.
Sports facilities and recreation: marina, sailing, underwater fishing, water-skiing.
Boat services: a bridge connects the island with the mainland: the point of access is from the Adriatic Highway, 27 km from Šibenik, 60 km from Zadar.

PAG

Towns and villages: Pag, Novalja, Stara Novalja, Lun, Povljana.
Accommodation: in a hotel and in private homes. Several thousand persons can be accommodated on campsites.
Sports facilities, recreation, entertainment: pleasure craft may be anchored independently, sailing, underwater fishing, water-skiing, naturist beach.
Boat services: ferryboat lines Jablanac-Stara Novalja, Stara Novalja-Pudarica (Rab), Karlobag-Pag, Karlobag-Metajna.

PALAGRUŽA

No possibility of accommodation.

RAB

Towns and villages: Rab, Lopar, Suha Punta, Supetarska Draga, Banjol, Barbat, Kampor.
Accommodation: in hotels and in private homes. Several thousand persons can be accommodated on campsites.
Sports facilities, recreation, entertainment: marina, indoor and outdoor pools, water-skiing, sailing, underwater fishing, mini-golf, bowling, tennis.
Boat services: ferryboat lines Senj-Lopar, Baška (Krk)-Lopar, Jablanac-Pudarica.

SUSAK

Accommodation: in private homes.
Sports facilities, recreation: sailing, fishing.
Boat services: small steamers at least three times weekly from the island of Cres.

ŠIPAN

Villages: Šipanska Luka, Sudjuradj.
Accommodation: in a hotel and in private homes. A campsite.
Sports facilities, recreation: sailing, underwater fishing.
Boat services: steamer services from Dubrovnik daily except Wednesday and Sunday.

VIS

The places on this island are not yet equipped to accommodate tourists, and information about day trips can be obtained in travel agencies on the mainland or in larger towns on the islands.

Prepared by Predrag Djuričić

PHOTOGRAPHY CREDITS

Colour-photographs

«Yugoslav Review» photography archives [photos by Dimitrije Manolev]: 14, 15, 17, 19a, 29, 30, 31, 34, 35a, 35b, 36, 54, 67a, 67b, 70a, 70b, 72, 73, 74, 75c, 75d, 76, 85, 86, 87, 89a, 92, 106, 107a, 107b, 107c, 108, 133, 134, 136, 137b, 138, 142, 153, 156b, 157, 158d, 163a, 166b, 167, 168.

Ivo Eterović: 13, 19b, 49, 51, 52, 53a, 53b, 66, 71, 75a, 75b, 101, 117, 118/119, 120/121, 122, 123, 124, 137a, 139b, 140/141, 154, 155, 156a, 158b, 158c, 159, 160, 161, 162, 177, 182/183, 184, 185, 186/187.

Milan Pavić: 32a, 32b, 47a, 50a, 104, 105a, 105b, 143a, 143b, 144, 158a, 163b.

Vilko Zuber: 4, 55b, 88, 89b, 188.

Mladen Grčević: 50b, 56, 90/91.

Turistička štampa, Beograd: 178/179, 180, 181.

All other photographs are from the Yugoslav Review photography archives.

Black-and-white photograps

Dragoljub Kažić: 24, 59, 79, 131, 146, 192.

Mladen Grčević: 96, 149, 150.

Maritime Museum, Dubrovnik [photos by Antun Tasovac]: 40, 173, 175, 196.

All other photographs are from the Yugoslav Review photography archives.

188. *A Gothic polyptych in the Franciscan monastery on the island of Ugljan, the work of a 15th-century native artist.*

Endpaper

This decorative map was charted in Amsterdam in 1689 and represents the territories under Hungarian sovereignty.

Maximæque Partis
DANUBII FLUMINIS
una cum adjacentibus et finitimis
REGIONIBUS
Novissima Delineatio
per
Carolum Allard
Amstelo-Batavum
OBERN
KRAIN
VNTER
KRAIN.
WINDISCH
MARCK.
FORUM
IULY.
ISTRIA
Gratz
Marchburg
Clagenfurt
Zagrabia
Triest
Aquilea
Grado
Venetia
Padua
Ferrara
Bologna
Carlstat
Idria
Goritia
Pola
Parenzo
Cherso
Arbe
Promontore
Apennini